Effective Science Teaching

DEVELOPING SCIENCE AND TECHNOLOGY EDUCATION

Series Editor: Brian Woolnough,
Department of Educational Studies, University of Oxford

Current titles:

John Eggleston: *Teaching Design and Technology*
Richard Gott and Sandra Duggan: *Investigative Work in the Science Curriculum*
David Layton: *Technology's Challenge to Science Education*
Keith Postlethwaite: *Differentiated Science Teaching*
Michael J. Reiss: *Science Education for a Pluralist Society*
Jon Scaife and Jerry Wellington: *Information Technology in Science and Technology Education*
Joan Solomon: *Teaching Science, Technology and Society*
Clive Sutton: *Words, Science and Learning*
Brian E. Woolnough: *Effective Science Teaching*

Effective Science Teaching

BRIAN E. WOOLNOUGH

Open University Press
Buckingham · Philadelphia

Open University Press
Celtic Court
22 Ballmoor
Buckingham
MK18 1XW

and
1900 Frost Road, Suite 101
Bristol, PA 19007, USA

First Published 1994

A catalogue record of this book is available from the British Library

0 335 19133 9 (pbk)   0 335 19134 7 (hbk)

Library of Congress Cataloging-in-Publication Data

Woolnough, Brian E.
     Effective science teaching/Brian E. Woolnough.
        p.   cm. — (Developing science and technology education)
     Includes bibliographical references and index.
     ISBN 0–335–19134–7.   ISBN 0–335–19133–9 (pbk.)
     1. Science – Study and teaching.   2. Engineering – Study and
teaching.   3. Science projects.   4. Curriculum planning – Great
Britain.   I. Title.   II. Series.
     Q181.W84   1994
     507.1 – dc20   93–41247
                                                                          CIP

Typeset by Type Study, Scarborough
Printed in Great Britain by St Edmundsbury Press, Bury St Edmunds, Suffolk

Contents

Series editor's preface

It may seem surprising that after three decades of curriculum innovation, and with the increasing provision of a centralised National Curriculum, that it is felt necessary to produce a series of books which encourage teachers and curriculum developers to continue to rethink how science and technology should be taught in schools. But teaching can never be merely the 'delivery' of someone else's 'given' curriculum. It is essentially a personal and professional business in which lively, thinking, enthusiastic teachers continue to analyse their own activities and mediate the curriculum framework to their students. If teachers ever cease to be critical of what they are doing, then their teaching, and their students' learning, will become sterile.

There are still important questions which need to be addressed, questions which remain fundamental but the answers to which may vary according to the social conditions and educational priorities at a particular time.

What is the justification for teaching science and technology in our schools? For educational or vocational reasons? Providing science and technology for all, for future educated citizens, or to provide adequately prepared and motivated students to fulfil the industrial needs of the country? Will the same type of curriculum satisfactorily meet both needs or do we need a differentiated curriculum? In the past it has too readily been assumed that one type of science will meet all needs.

What should be the nature of science and technology in schools? It will need to develop both the methods and the content of the subject, the way a scientist or engineer works and the appropriate knowledge and understanding, but what is the relationship between the two? How does the student's explicit knowledge relate to investigational skill, how important is the student's tacit knowledge? In the past the holistic nature of scientific activity and the importance of affective factors such as commitment and enjoyment have been seriously undervalued in relation to the student's success.

And, of particular concern to this series, what is the relationship between science and technology? In some countries the scientific nature of technology and the technological aspects of science make the subjects a natural continuum. In others the curriculum structures have separated the two, leaving the teachers to develop appropriate links. Underlying this series is the belief that science and technology have an important interdependence and thus many of the books will be appropriate to teachers of both science and technology.

In this book I have reconsidered what makes effective science teaching. Based on my recent research, which looked into the factors that affect a student's decision towards or away from science and engineering, and my own experience of school teaching and teacher education over the years, I have become increasingly convinced of the need to find ways of making science in schools a holistic activity. I am using the word 'holistic' in a double sense, both of incorporating the affective as well as the cognitive and psychomotor areas of a student's life and also in the type of scientific activity in which the students are involved. Too often in school science we have produced a reductionist model of science, reducing it to small, disconnected pieces of knowledge and trivial practical exercises unrelated to genuine scientific activity. In this book, I argue that at the heart of effective science teaching is the active involvement of students in scientific research projects. I argue, and illustrate from practice in schools, that this is not only possible but also effective, both in helping students to enjoy and mature in personal and scientific autonomy, and also in encouraging many students to continue with science or technology into their adult lives.

Brian E. Woolnough

Acknowledgements

In England and Wales, as in many other countries, science is not as popular as it should be. Too few students are choosing to continue with their science into higher education and a career, and too few appear to enjoy the subject while they are studying it in school. And yet, for many of us and for many students too, being involved with science is a fascinating, stimulating and highly enjoyable activity. Why don't all students find it so?

Science in schools is often criticised, especially by older students, for being too prescribed, too impersonal, too lacking in opportunity for personal judgement and creativity. Science has become reduced to a series of small, apparently trivial activities and pieces of knowledge unrelated to the world in which the students are growing up and inhibiting their developing personalities and aspirations. And yet, despite the pressures of National Curriculum and examination syllabuses, science in schools can, and often is, much more exciting and stimulating. This book celebrates such science teaching, and the qualities of the teachers who produce it, and argues that doing science should be a holistic and not a reductionist activity. It should involve the affective as well as the cognitive aspects of a student's life. It is not sufficient to be concerned only with what students know and can do; one must also be concerned with whether they want to do it. It is of fundamental importance to develop students' emotional involvement with their work; to develop their motivation, their commitment, their enjoyment and creativity in science – for without these any knowledge and skill they acquire in the subject will be of no avail.

Furthermore, it argues that the best form of effective science teaching is through student research projects, in which students take a problem of personal concern to themselves and tackle it, worry at it, persevere in it and, meeting its challenges, produce their own solution. Such involvement in genuine scientific activity is, I believe, not only possible in schools but also essential if school science is to do justice to our students and to the scientific enterprise itself.

In writing of these matters (the importance of a holistic approach to learning science especially through student research projects) I am conscious that they have been recurring themes throughout my time in science teaching. During this time I have gained much encouragement and inspiration from many people, to whom I gratefully acknowledge my debt.

Of prime importance are the lessons I have learnt from the many students with whom I have worked; they have shown me through their own performances in doing projects and investigations the high standard of genuine scientific work of which young people are capable. I have learnt that we rarely overestimate their capability, and that the more responsibility we encourage them to take for their own learning the more their enormous potential will develop and the more they will appreciate, enjoy and mature in science. I also acknowledge the advice, ideas and encouragement I have received from untold teachers who have shared with me their experience of project work in science.

In particular, I acknowledge the inspiration I

received from Bill Bolton who many years ago, while teaching at a college of further education, spoke to a meeting of the Association for Science Education (ASE) about the projects he was doing with his students and who made me, as a newly appointed head of physics in a school, realise that this was the type of practical work I was looking for and thus 'switched me on' to student investigations in my teaching.

I acknowledge the example of Ted Adnams, a rural science teacher who devoted his life to teaching science to students in a local comprehensive school by involving them in genuine scientific work of the highest level; studying mallards and mice, establishing ponds and nature trails in the school grounds, and investigating many forms of animal and plant life. By involving students in such investigations he showed generations of young people what real science was about as they worked alongside a man who was himself a first-rate scientist, showing great rigour, respect and love for the world they were studying.

I also acknowledge with gratitude the Nuffield A-level physics examiners who allowed me the privilege of being a moderator for the A-level Investigations over a period of years. This gave me the opportunity of seeing again what impressive scientific work students were able to produce when given the opportunity. Of the hundreds of reports I saw, very few were disappointing, the vast majority were very good indeed, and many were exceptionally good – being models of scientific achievement of the highest order.

In the production of this book I acknowledge the many teachers and students whose experience, and permission to reproduce their work, have given reality to general principles. I have deliberately included lists and examples of a wide range of student investigational work, for I remember how necessary I found this when starting to do investigations with my own students.

I am grateful to the students and teachers who helped me with the FASSIPES research by sharing their views and experience. (FASSIPES (Factors Affecting Schools' Success in Producing Engineers and Scientists) was a research project funded by the Institute of Physics and the Institution of Electrical Engineers.)

I acknowledge with thanks permission to reproduce work in this book given to me by the following: the Institute of Physics for three of my editorials for *Physics Education, 'Reductio ad absurdum?'*, 'Investigations – not a "tame" type of practical!' and 'Travel can broaden the mind!'; and 'Project Superconductor' by Lynne Pebworth from John Mason School; Alan West for 'The baby's "Clearway" project'; the Association for Science Education for 'Controlled high speed photography of milk splashes' from the *School Science Review*; Alistair Ross for his 'The Strongest Conker'; the British Association for 'Raindrops keep falling on my head' from SCOPE, and the case study for the silver award of Young Investigators; CREST for their Silver Award Profile, and for involving me in their most impressive and stimulating CREST awards ceremonies; ICI, and Ken Gadd from Yeovil College, for his 'From artificial kidneys to novel catalysts'; the Engineering Council for Neighbourhood Engineers in Action and Barbara Zamorski's 'Four Schools in Devon and Cornwall'; Russell Tytler for quotations from his Australian study; Brian Ford for 'A heretic confesses'; Mr J. Campbell and Nigel Connor, Ryan Doherty and Gary McIntyre from Coleraine Boys Secondary School for 'Abrasion damage to leaf surfaces'; Mr Gordon Woods and Peter Glynne-Jones from Monmouth School for 'Bays Masterminds Science Quiz'.

I am grateful to those who have provided me with additional photographic illustrations and the permission to reproduce them, and to the (unacknowledged) subjects of these photographs who, through their enthusiasm and creativity in these projects, have done so much to give personal reality to my impersonal words. To CREST for Figures: 4.1, 4.3, 4.4 and 5.5; to Jodrell Bank Science Centre for Figures 5.3 and 5.4; to Guy Molyneux for Figure 5.6 and the Royal Institution for Figure 5.1.

I am grateful to all these, and wish readers much joy and satisfaction with their students in doing effective science teaching.

CHAPTER 1

Aims and unresolved tensions

Introduction

This book is a celebration of good practice. It is an affirmation of that type of science teaching that has been so effective in influencing young people to enjoy and continue with science through school and into adult life. Above all, it is dedicated to those innumerable science teachers who have, through their enthusiasm, expertise and commitment beyond the call of duty, made science such a stimulating and worthwhile pursuit.

It has as its central theme the vital importance of extra-curricular activities in science. What goes on in science lessons, the bread and butter of science teaching is undoubtedly important. But many students are really 'switched on' to science and given personal satisfaction in it by 'the little extra bits' over and above the basic syllabus. For some it is the inspirational enthusiasm and personal encouragement of an individual teacher. For others it is involvement in a science competition or a great egg race. Yet others find it in a spectacular lecture demonstration, a stimulating visit or the involvement of a local engineer. And for all it is the sense of personal involvement, commitment and autonomy in doing science which make it so stimulating and effective.

There is more to learning science than gaining knowledge and developing skills. The area of the affective, the personal acceptance and enjoyment of and commitment to the scientific activity, is of central importance in doing science. Doing science, in school or at work, is a holistic activity involving the will and the imagination as well as the mind and the hand. It should not be reduced to fragments of knowledge and skills. I hope that the discussion and examples given in this book will help to develop such real scientific activity further, and make science teaching not only more effective but also more fun!

A variety of aims

One of the complications in teaching science is that it has no single, unambiguous aim. Different groups in society, different teachers, have different aims. Some see the aims in vocational terms, in providing a suitably skilled workforce for developing industry and the national economy, others see the aims more in educational terms, in developing the individual potential of each child. Some see the aims as transferring knowledge, others in developing skills or attitudes. Some aim to produce highly qualified potential scientists and engineers, others aim to produce a broad, well-educated citizenry. Furthermore, the aims that are appropriate to one student may not be appropriate to another: the aims appropriate for an education *in* science for those with a 'scientific turn of mind' may not be equally suitable for an education *through* science for those not wishing or able to continue with their science studies beyond 16. Sometimes, but not always, a certain type of teaching will satisfy more than one set of aims (Fensham, 1985).

One of the challenges in teaching science is to

ensure that the most appropriate aims are met for all students, and to ensure that in seeking to fulfil one set of aims for some students we do not hinder the realisation of other aims for other students. In much of the curriculum development of the past we have been too optimistic in believing that the same type of science is appropriate to meet every goal. It will certainly help if we first clarify some of the important aims, and then we can more readily see how to meet them.

There has in the past been no shortage of statements of aims for science teaching. Science educators, scientists, politicians and industrialists have delighted in declaring them. Often these have then been taken forward into more specific objectives, sometimes specified as behavioural or assessment objectives. Examination syllabuses and national curricula are full of them. They are quite specific, they can be translated into teaching, they can be measured reliably, but in themselves they may be far removed from genuine scientific activity. The danger is that, when we transfer from broad aims to specific, measurable objectives, we lose some of the most important aims and reduce science to an accumulation of trivial fragments. Eric Rogers (1977) saw the danger, sadly not recognised by much subsequent curriculum development, when he said:

> It is fashionable in Europe now to carry out a meticulous analysis of separate objectives and outcomes of teaching and learning, so that they can be assessed in tests. This 'taxonomy of educational objectives' grew in the work of Bloom and others in the United States twenty years ago. As it developed it was a valuable revolt against carelessness, vague planning and testing. But it concentrates attention on aspects that are clearly measurable and it misses some of the most important factors in our hopes for lasting benefits from Nuffield science – the enjoyment, ambition, pride, that we look for in 'wonder and delight and intellectual satisfaction'.

A reductionist approach to the science curriculum has done much to reduce the validity of the whole scientific enterprise. The whole is more than the sum of the parts, and different from it (Woolnough, 1989). The thrust of this book is that we

should do all we can to maintain the wholeness of scientific activity, that students should be continually given the opportunity to do genuine scientific investigations, and that we should beware of reducing science to a series of pseudo-scientific exercises, be they practical or theoretical.

Rather than looking at a series of second-hand aims for science teaching, science teachers working together might find it more profitable to write down, discuss and develop a few of their own aims for their science department. I still find it helpful to think in terms of education *through* science and education *in* science, and to consider how far these are appropriate for our different students. Education *through* science uses science lessons as the vehicle for teachers to further general educational aims such as interpersonal skills, self-confidence, even an awareness of the significance of science in society; education *in* science is concerned with learning about the specific content and processes of science itself. In the past a false linkage has often been made, associating education *through* science with the less able, or less motivated, students and education *in* science with the more able. I would suggest that both are applicable, to a greater or lesser extent, to all students, whether or not they are destined to continue with science after school. Both are appropriate for science for citizens and science for scientists. Box 1.1 summarises the variety of aims which we might want our students to achieve.

In many ways education *through* science is a subset of education *in* science, and should not be equated with it. Many of the so-called processes of science are merely general life skills which, while important in themselves, do not equate with the way scientists work. Problem-solving skills are context-dependent and scientific problem-solving needs an understanding of the relevant scientific context. The knowledge of science which is useful for, say, responsible household management and gardening is not the same as the scientific knowledge of theories and models which so delights and stimulates the scientist, nor the same as the social implications of science which will be important for all citizens. It is in this area of the content of science courses that much potential conflict arises.

Box 1.1   *Aims of education* through *science and education* in *science*

Education *through* science	● Attitudes	● Self-confidence, pride in work ● Autonomy and commitment ● Integrity in thought, presentation and debate
	● Skills	● Communication skills – literacy – oracy – numeracy, including IT ● General problem-solving skills ● Co-operation and other interpersonal skills
	● Knowledge	● Useful scientific facts ● Knowledge, understanding and appreciation of our world
Education *in* science	● Attitudes	● Enthusiasm for science, wonder at physical and biological world ● Humility about limitations of science
	● Skills	● Use of scientific apparatus ● Problem-solving in scientific context ● Scientific data analysis and reporting
	● Knowledge and understanding	● Knowledge of the important facts and theories of physical, biological and earth sciences ● Understanding and appreciation of scientific facts, theories and models ● Ability to use scientific knowledge to solve problems

There are some important areas of science which, while not useful in themselves, are part of our cultural heritage. There are some scientific concepts and techniques, some of the more abstract and mathematical ones, which, while being inappropriate for many students, will be both stimulating and motivating to others.

All the aims mentioned above have been focused upon the needs of the individual student. There is another perspective, more commonly held by those outside the educational system than by those in it, which relates to the needs of society. While not denying the importance of any of the above personal aims, society (represented by politicians, industrialists, people in business and the service industries, the media and parents) also focuses on the need to gain good qualifications, acquire useful attitudes and skills, and be prepared to obtain appropriate employment. Politicians and industrialists see one important aim for the school system as the production of an adequate supply of students for those jobs which will support and regenerate the national economy. They expect all

science departments to have 'producing an adequate supply of students heading for careers in science and engineering' as one of their aims, though perhaps not the only one.

So we have a variety of ways of defining 'effective' science teaching, corresponding to the variety of aims. One way undoubtedly takes in those methods of teaching science which encourage students to continue into scientific careers or engineering. But that would not be appropriate for all students. Other measures of effective science teaching are that students should enjoy their science and thus obtain a better appreciation of the wonderful world in which they live and their place in it (though, like so many of the really important things in life, this is immeasurable!). Further, they should know and understand some of the principles and facts of science, that they might be scientifically literate and appreciate both the cultural and the useful aspects of science. We would expect them to have appreciated the effect that science and technology have on society, and that society has on science and technology, and to realise that they themselves can have some influence on it. We would also expect effective science teaching to leave students with the ability, and the motivation, to solve problems scientifically and to carry out scientific investigations, and we would expect them to have developed their self-confidence, their ability to work honestly and cooperatively, and their ability to communicate the results of their work through reports and presentations. And it is here, in these general, educational goals that we can best meet the needs of society as well as of the individual. For it is just these achievements – of self-confidence, of personal and communication skills, and of autonomous, creative, well-grounded working habits – that will best meet the needs of industry and business, and produce an educated citizenry.

Unresolved tensions

In looking at ways of making science teaching more effective, it will be necessary first to consider the present situation and analyse some of the unresolved issues. Though some see the present situation of science teaching as a disaster (Claxton, 1991), full of problems to be solved, I find it more helpful to consider it an imperfect but healthy organism, which has developed as the result of many conflicting influences and which now contains many, often inevitable, tensions. Tensions can be creative, problems are usually depressing! We can then see how these tensions can be accommodated, ameliorated or harnessed (or side-tracked!). Throughout this book we will be considering how such issues arose (in understanding how we got here we can better decide about alternative routes), we will look at some research into students' and teachers' reactions to current practice (in basing our discussion on broad evidence we will gain greater insight than individual experience or anecdote would provide), we will consider examples of proven good practice (and examine why it is so effective) and finally we will consider ways forward for science teaching in schools (that the problems might be lessened and the successes increased).

It might seem surprising that, in countries with an established national curriculum 'given' to science teachers to be 'delivered', there is still a need to reconsider issues relating to the science curriculum itself. But such national curricula are, inevitably, pragmatic solutions resulting from compromises between differing vested interest (Woolnough, 1988; Dobson, 1992). Ultimately it is teachers who mediate the curriculum to the pupils, who determine how it is presented and how students experience science. The curriculum provides a framework within which science teaching can flourish, not a strait-jacket to restrict it. The science curriculum is best seen as a living plant to be tended, not a sterile package to be delivered (Woolnough, 1990a). In England and Wales there are, inevitably, still tensions within it, and science teachers need to harness them and reconcile them in ways that best meet their own vision and expertise, as well as the needs of their students. There is still much that needs to be considered in the way the National Curriculum is to be taught. We will consider the tensions under two broad headings: those relating to the formal curriculum itself; and those relating to its organisation and manner of delivery. Both will, in some respects, also relate to the all-important 'hidden' and informal curriculum.

Curriculum issues

Continuity

The arguments for continuity in the science curriculum have been strong. Students should know that when they move from one class to another, or from one school to another, they are not going to repeat material or miss out important topics. If science is important enough as a subject to be included in the National Curriculum then it should be possible to make explicit the nature of the subject, both its content and its way of working. We cannot argue that it is essential that every student should study science, and that it is up to individual teachers as to what they mean by science. One of the problems of primary school science teaching in England and Wales in the past has been the considerable variation in what has been taught. There has been variation in the topics taught and in the overall approach: whether science should be taught primarily as a body of knowledge or as a process of investigation. Though much excellent science has been taught, depending largely on the individual teacher, the lack of consensus has meant that secondary schools receiving students from a number of primary schools have had no common framework on which to build.

But while the emphasis on a common framework within a national curriculum has obvious strengths, the consequent imposition of too rigid a structure can cause the removal of the creativity, flair and commitment often associated with teachers teaching to their own particular strengths and enthusiasms. This has been a real loss. With a very heavy content load in many national curricula, and only that content being assessed, it is very easy to find the individuality forced out. The tradition in England and Wales of the teacher as an autonomous professional does not respond well to the 'delivery' of a given curriculum, which fits more the model of 'teacher as channel'. The feeling of being deskilled which has come with the imposition of the National Curriculum is not in anyone's interest.

It will be argued in this book that by using projects and investigations, it is quite possible to have the best of both worlds. Teachers' and students' individuality and flair can coexist with a given structured curriculum, and can be mutually supportive. Indeed, personal creativity and knowledge are mutually dependent and cannot develop satisfactorily without each other.

Progression

The idea of progression within a spiral curriculum is, again, self-evident. If a topic is covered more than once in a student's career we would expect that it is done at greater depth at a later stage. We would expect 16-year-old students doing practical investigations to demonstrate a more mature approach than they did at age 11. But the nature of progression in science is often not linear; it is more like a roller-coaster ride than a climb up a regular flight of steps. Thus tensions arise when we pretend that we can predict what progression means, particularly when we have to describe that progression in ten specific stages for every student from the age of 5 to 16, as has been decreed for the National Curriculum in England and Wales. Such a prescribed progression of criterion referencing is often spurious, and should be recognised as such. Those of us who worked on the DES Grade Related Criterion working parties in the different subjects realised the impossibility of the task back in 1987: it often looks meaningful in print but lacks practical reality. In practice, science teachers have to give meaning to the statements by norm-referencing from their experience. It is true that some topics are generally found to be more difficult than others – compare, for example, the replication of DNA and the parts of a body; understanding chemical reactions in terms of energy transfers and the processes rusting and burning; electromagnetism and current electricity. Abstract concepts are more difficult than concrete ones. But even with the amount of research that has been done on the difficulty of different topics by the Assessment of Performance Unit (APU) and others there is insufficient evidence for differentiating into ten different levels. The problem is compounded when we move from 'knowing that' or 'knowing how' to 'understanding'. What level of understanding is expected? I suspect that questions could easily be set which enabled us to demonstrate that we did understand or that we did not understand a particular principle, depending how difficult the questions were. To take two examples from the National Curriculum for science, do we really 'understand the life processes in green plants' (how does sap rise up a tree more than 10 metres tall?) or 'understand the nature of electromagnetic radiation in the process of interference' (what if we detect the interference patterns with photoelectric cells?)?

The problems with prescribing progression in ten tidy steps are even more considerable when related to the ability to do investigations. Teachers can certainly make *post-hoc* judgements about the quality of investigations after they have been done, measuring against predetermined criteria (as has been done successfully by physics teachers in Nuffield A-level physics for over twenty years). But it is not possible to prescribe how a particular investigation might go. Normally, controlling a number of variables is more difficult than controlling one; quantitative investigations are usually more appropriate than qualitative ones. It is not possible to define grade related criteria meaningfully and unambiguously. It is possible to perform a difficult investigation splendidly with only one variable; in some cases a qualitative investigation is more appropriate than a quantitative one. The ability to carry out a good scientific investigation, in school or in the laboratory of a research scientist, often depends more on commitment, flair and tacit knowledge than on technical skills and explicit knowledge of scientific theory.

We will see that while progression is important, it needs to be handled sensitively and lightly, relying more on the professional judgement of teachers and other experts than on legalistic formulations of prescribed stages. General criteria are useful, but they only acquire meaning as they are norm-referenced by professionals. This can be done both in school assessment and through extra-curricular activities, as the assessment schemes for CREST awards and egg races (see Chapter 4) have shown.

Differentiation

Any class is likely to have pupils with a wide range of ability, motivation and career potential. It would be good to be able to differentiate the curriculum so as to provide science experiences appropriate to each individual student, so that all can fulfil their potential. Some students find the concepts of science easy and fascinating, while others find them dull and difficult, especially where mathematics is involved. But providing a differentiated curriculum is not easy. One alternative is to group students by ability and thus make

different curriculum provisions for different groups. Grouping pupils by ability necessitates crude discriminating techniques which ignore many student variables, encourages sink forms and underachievement, and acts contrary to many teachers' instinctive desire to treat all children equally. Many teachers would want to emphasise what students have in common, their common entitlement and their common potential, rather than to emphasise and thus increase their differences. A second alternative is to keep mixed-ability groupings but to offer alternative experiences, possibly through individualised learning, within each class. Though not insurmountable, the organisational difficulties of this approach are enormous and often result in all students following unstimulating courses, even if they are doing so at their own pace. The tension between differentiation and commonality is considerable. I believe that in recent years many schools have overemphasised the common entitlement of all students and have failed to differentiate appropriately. Previously we had tended to overemphasise the differences and thus failed to recognise and so increase all the potential that students had in common!

We will show that much headway can be made in differentiating appropriately by teaching a common curriculum to different depths in broad-ability groupings, and providing more open-ended tasks throughout the science curriculum to enable students to respond differentially. Student research projects allow 'differentiation by outcome', with stimulus and satisfaction for all.

Accessibility

Much development has taken place since the 1980s to make the science curriculum more accessible to an increasing range of students. Textbooks have become much more personal and user-friendly, and teaching schemes have built on students' own experience. This has made the sciences much more popular with the majority of students, with both boys and girls feeling that it is relevant to their lives. Much of the older impersonal, abstract, conceptual science has been omitted, along with many of the mathematical applications which students were finding increasingly difficult. But in making the sciences more accessible we have dropped (particularly from the physical sciences) some of the rigour, the precision and the 'elegance' of the concepts which some of the brighter students found so stimulating. There is a tension between making on the one hand, an accessible 'science for all' and, on the other, an intellectually rigorous 'science for future scientists'. Nevertheless, I hope to show that it is possible to provide a basic curriculum which is accessible and satisfying to all, and yet still to include the necessary conceptual and mathematical stimulus for those who appreciate it.

Process or content?

The debate about the balance between process and content in the school science curriculum has been running since science was first taught in schools (Layton, 1973; Jenkins, 1979). To what extent should science be seen as a corpus of knowledge, and science teaching as the transfer of culture? To what extent should science be seen as a way of working, and science teaching as enabling students to learn the methods of science? Few would doubt that school science should contain something for both. But that does not resolve the argument. What, and how much, content? What, and how much, process?

Of the two, the question of content is (slightly!) less contentious as long as we realise that a selection will have to be made and that such a selection will, to some extent, be arbitrary. The following criteria for selection of scientific content could be made. (Not all items of content included will fulfil all of these criteria, but they should at least satisfy some!)

Any scientific topic included should:

- be significant science, recognised by scientists as being among the central aspects of their subject. Students would not be competent in science if they did not have a knowledge of, say, how plants and animals function, how chemicals react, how particles respond to forces.

- present a coherent picture of science, emphasising some of its unifying principles. It would be easy to offer a rag-bag of apparently unconnected items, but this would fail to give a glimpse of the intellectual elegance of the subject.
- be comprehensible to the student at that stage. We know that abstract concepts cannot be fully appreciated by young students still thinking concretely, but we are also learning not to underestimate students' ability to comprehend difficult ideas if they are presented in appropriate ways and build on the students' imagination and prior knowledge.
- be useful to students both in everyday life and as a foundation for further study. There are many aspects of science which students need to know – about health and nutrition, acidity of soils, household wiring – and these clearly have a place in a school science course. Some students will want to follow science into further and higher education, or into employment, and note should also be taken of what is useful and necessary for that.
- be relevant and interesting, taking relevance in its wider sense. The recent successful inclusion of astronomy and geology in the National Curriculum for England and Wales illustrates the relevance that an understanding of the wider world around us can give to students. Similarly, the broadening of science content to include 'the application and economic, social and technological implications of science', to use the words of the National Curriculum for science (DES, 1991), demonstrates the need to make science relevant to students' lives (Solomon, 1993).
- be intellectually stimulating. There is evidence that many students are switched on to science by the intellectual stimulation gained from some, often esoteric, topic in the subject. The elegance and challenge of genetic programming, of chemical energetics, of cosmology and relativity (Kalmus, 1985) have fascinated and provided the incentive for many able students to continue their studies into higher education and a career in science.

The debate about the nature of the appropriate process(es) of science is far less straightforward. We can analyse the way a scientist works and separate out some of the component skills, and this has been done in many curriculum and assessment projects such as Warwick Process Science (Screen, 1986), Techniques for Assessment of Practical Skills (Bryce *et al.*, 1983), the Graded Assessment in Science Project (Swain, 1989) and the GCSE examinations. But such a reductionist approach is not very helpful: doing science is more than being competent in a series of scientific skills. The whole activity of doing science does not equal the sum of the parts, it differs from and exceeds it (Woolnough, 1989). Furthermore, many of the processes that a scientist uses, such as planning, hypothesising, observing, measuring, inferring and reporting, are in fact general life skills and very much context-dependent (Millar, 1991). They only become scientific processes when set in the context of a scientific activity and interpreted with scientific understanding. Observing as many things as possible about a situation is not good science; it only becomes so when the scientifically significant things are discerned. So, the processes of science can only be developed alongside scientific understanding and in the context of doing a scientific activity, such as an investigation. This is now stressed in the National Curriculum for England and Wales, which introduces its first Attainment Target on Scientific Investigation thus:

> Pupils should develop the intellectual and practical skills which will allow them to explore and investigate the world of science and develop a fuller understanding of scientific phenomena, the nature of the theories explaining these and the procedures of scientific investigations. This should take place through activities that require a progressively more systematic and quantified approach which develops and draws upon an increasing knowledge and understanding of science. The activities should encourage the ability to plan and carry out investigations in which pupils:
> i) ask questions, predict and hypothesise;
> ii) observe, measure and manipulate variables;
> iii) interpret their results and evaluate scientific evidence.

The assessment of a student's progress in this must be

> carried out in the context of a scientific investigation. Discrete practical tests that lead to the assessment in isolation of specific practical skills . . . do not fulfil the requirements. (SEAC, 1992)

There are two other processes relevant to the scientific exercise. We have been thinking above about how a scientist works practically. But scientists, especially but not exclusively physical scientists, also work with conceptual models, with language and mathematics. Much of the process of scientific thinking is not done in practical work but in making sense – through language, models and analogies – of what is observed. These issues are discussed well in Sutton (1992). The way scientists solve problems quantitatively, using mathematics, making approximations and 'back of envelope' calculations, is another important process of science and should be developed, more than it is at present, through school science courses.

It is a central thesis of this book that there is a need to develop students' understanding and application of a serious body of scientific knowledge and that the 'processes' of science, and their symbiotic relationship with theory, can best be developed through student involvement with whole projects, competitions and scientific investigations.

Assessment

Student assessment in science is a necessary and potentially beneficial aspect of science teaching. It provides feedback to teachers, students and parents concerning students' achievement. It can be both formative, to diagnose and help develop succeeding learning experiences appropriately, and it can be summative, to assess achievement and grade, to direct or select students as required. The danger with assessment is that only certain things are assessable and those things tend to acquire undue importance because only they are assessed and reported. We can assess a student's knowledge of science and its application to certain types of problem, but we cannot assess a student's commitment to or enjoyment of science.

There is a further tension between reliability and validity in assessment, particularly relating to students' ability to do practical science. Because of the public currency of many forms of assessment, people rightly demand that the assessment should be reliable – which is to say that the result of the assessment should be reproducible no matter who the assessor is. But an excessive emphasis on reliability can lead to assessment of things which have low validity – things which do not usefully reflect the important aspects of students' achievement in science. Some of the more important aspects of ability in sciences are not susceptible to tight marking and need the professional judgement of experts. In gaining validity we may have to be prepared to lose a little reliability. When a student has done an extended, practical investigation it will not be easy to award it a mark with absolute reliability, for there will be idiosyncratic, unpredictable paths which the student will have followed and there will be particular areas of strength and weakness. We could obtain a more reliable measure of the student's ability to read a thermometer accurately, or to record results appropriately on a graph, but this would not be a valid measure of the student's ability to do a real science investigation. Nor is it possible to give genuinely reliable assessment in anything other than the most trivial, atomistic, assessment exercise. Box 1.2 amplifies this argument.

In attempting to make our assessment more reliable we have been forced to emphasise the easily measurable, and are in danger of letting the tail of assessment wag the dog of science activity in a destructive way. Constructive compromises have been achieved by a mixture of norm and criterion referencing, relying on the professional judgement of teachers, supplemented by a sensitive system of moderating. Perhaps the best example of this was the old Nuffield A-level physics assessment of the long practical investigation. In this, general criteria for a good scientific investigation were spelt out and teachers assessed students' progress and final achievements according to how well they met these criteria. An external moderator moderated the grades given on the basis of students' written reports and the teacher's comments (Jakeways, 1986).

We hope to show that the assessment of real science investigations can be done painlessly and constructively by using professional judgement against broad criteria, as in schemes such as CREST and Young Investigators (see Chapter 4).

Girls and science

While it is true that each individual student is different and reacts differently to different approaches to science, there is a large body of evidence that suggests that girls, on the whole, react to some aspects of science teaching differently from boys. Girls, especially through the formative adolescent years, show an aversion to the physical sciences, particularly physics. Boys often seem less keen on the biological sciences. These preferences are shown up when students are given a choice of subject for further study, whether at the age of 14, 16 or 18. In most, though not all, countries there will be a preponderance of boys in the physics classes and a preponderance of girls in biology. Society, for interesting sociological reasons which we will not discuss here, considers the former situation more cause for concern than the latter.

What is it about the sciences, and the way they are presented in school, that causes boys and girls to react to them differently? Much research has been done to seek answers to this question (see, for example, Kelly, 1987; Sjoberg and Imsen, 1988; Head and Ramsden, 1990; Murphy, 1991), and the main consensus seems to focus on social conditioning largely, but not exclusively, outside school and on the nature of science as presented to students in school. In society at large, from the time they are born, girls are encouraged to believe that science or engineering is not for them, unless it is one of the 'soft' sciences to do with people, animals or flowers: medicine or biology. The pervasive messages of role modelling, of parental expectations, of toys given (dolls not mechanical

Box 1.2   Brian Woolnough, 'Reductio ad absurdum?', editorial in Physics Education, 23 (1988)

Few outside the English educational scene will be able to understand why there is so much anguished concern about the UK government's recent proposal to introduce a National Curriculum. It is generally accepted in most countries that what is taught in schools is determined centrally, and the regular assessment of pupil achievement by standardised tests is an accepted part of life in many countries: and still life goes on! In England and Wales, however, there has been a long and cherished tradition of teacher autonomy concerning what is taught, the only arbiter and determinant of a relative commonality among the schools being the national school examination for 16- and 18-year olds. This lack of central control has encouraged teachers to innovate and experiment; it is doubtful whether the Nuffield physics courses would have developed in a more centralised system.

The most fundamental concern centres around the proposal that all pupils should be assessed and tested, in a standardised manner, at the ages of seven, 11, 14 and 16. The more experience we have gained over the past few years concerning the assessment of pupil performance in science, and the more we have thought about what constitutes the true nature of scientific activity, the less confident we have become that we can assess a pupil's ability to do science with any degree of validity. We have also become aware that the very act of assessing practical work according to a tight pre-determined schedule can significantly distort and damage real scientific activity by the student. An extension of Heisenberg's uncertainty principle perhaps!

There has been in the UK recently, as in other countries, a greater stress on the processes and skills of science. In analysing the nature of scientific activity, emphasis is given to such broad processes as planning, observing, manipulating apparatus, recording and interpreting data, pattern finding and reporting and communicating results. Going further still, specification is made of particular skills that a pupil should acquire, such as the ability to use a thermometer or to read an ammeter. Such analysis is entirely appropriate as an aid to diagnosing student weaknesses and encouraging particular aspects of science teaching: the problems arise when we try to tie precise assessment techniques to such processes and skills.

As we endeavour to assign specific criteria to such skills and processes we find ourselves moving into untenable paradoxes: either we define very precise criteria which become trivial or propose broad general criteria which become relative. If we want to make a statement, for instance, about a student's ability to use scientific apparatus and produce such criterion statements as 'uses measuring instruments correctly', we beg the questions about the type of measuring instrument, the context in which it is being used and the meaning of the word 'correctly'. So we try to make our statements more precise, but then find that even apparently small variations in the conditions can make very large differences in performance. The Assessment of Performance Unit (APU) found, for instance, that whereas 61% of pupils could correctly read a voltmeter with main divisions marked in 1 V steps and subdivisions of 0.1, only 11% of pupils could correctly read an ammeter with main divisions marked in 0.1 A steps and subdivisions of 0.02. Do we therefore need to write out criterion statements such as 'can read a voltmeter, not less than 100 mm long, with unitary divisions, with no zero error, if the light is good, after a good night's sleep, to an accuracy of ±0.2 V, more than 80% of the time'?! Clearly that would be silly and entirely inappropriate to the needs of pupil, teacher or employer. All that needs to be known is whether the pupil is relatively good at handling scientific apparatus. That in turn forces us back to a simpler, looser, more holistic form of assessment relying on the teacher's professional judgement. Such a form of assessment is quite out of sympathy with the government's aim of standardised testing of pupils for reasons of accountability. We have a problem!

At the heart of the problem lies the very nature of scientific activity. With the growing emphasis on processes and skills, we have been led to believe that being good at science is a matter of being good at the component processes and skills. We have produced a new reductionist model of scientific method comprising the compilation of certain predetermined parts, in the right order. But real scientists do not work like that. Real science is a much more messy, more human, more unpredictable process. Peter Medawar asks 'Why are most scientists completely indifferent to – even contemptuous of – scientific methodology?' and replies 'Because what passes for scientific methodology is a misrepresentation of what scientists do or ought to do'. He insists that scientific work proceeds by intuition, serendipity and imagination, followed by rigorous attempts at disproof. Percy Bridgeman asserts that 'The scientific method consists of doing one's damndest [sic] to understand nature, no holds barred'. Such a humane model of scientific activity can be introduced into school science teaching through the use of individual, problem-solving, practical investigations. Some teachers, especially through the practical investigations of Nuffield A-level Physics, have had experience of this for some years and have seen what excellent scientific work can be achieved by school students. More significantly, they have also shown that such investigations can be assessed holistically, without having to break down the analysis into specific, predetermined, processes and skills [see Jakeways (1986: 212)]. Let us hope that any assessment of scientific progress in the new National Curriculum, and throughout physics teaching, can be based on such constructive, sensitive and fundamentally perceptive methods.

kits), of what a woman should be like as depicted by the media and girls' magazines, are still strong and highly gender-specific. Many such messages are imprinted from childhood, into boys and girls alike, and discourage girls from taking science seriously. The stereotyping is strong, if not stronger, in boys than in girls. Indeed, in school it is the boys who are the bigger cause of the problem than the girls. At the age of 11, for instance, girls believe they can do science and technology as well as boys, but the boys believe that the girls are not so good. After a couple of years subject to such sexist attitudes among the boys, the girls begin themselves to believe that they are not as good as the boys at science and technology.

But if there is little that teachers can do to change such expectations promoted by society outside school, the school science curriculum and the way it is taught can be adapted to become more girl-friendly. The science content preferred by girls involves or relates to real people – health, growth and medical applications. Being socially more mature, they do not 'see the point' of many bits of dehumanised knowledge or theories, such as the mole concept or magnetic fields, unless they are shown to be related to the real world. Putting science into its human context has very positive effects on the attitudes of girls, apart from making it more effective and appropriate science for all students. Showing that science and technology are about solving real human problems gives science a sense of purpose.

The practical work that is often done in science – closed, decontextualised, apparently 'pointless' exercises that have to be tackled in a prescribed way – does not match the way many girls tackle problems. Patricia Murphy (1991) has illustrated how girls set about problems in a more explora-tory, more cooperative way, seeking more human-related solutions, and needing more time to reach a (better?) solution than the boys. She quotes the differences observed when boys and girls were asked to design boats that would travel around the world. The boys concentrated almost exclusively on the mechanical features of how the boat would float and be driven, and on accompanying weaponry. The girls were more concerned with designing living quarters and providing for the necessary food for the journey, none of which the boys considered. School science, if not the real world, tends to favour the boys' analytical ap-proach rather than the girls' more people-centred one. However, we hope to show that many of the investigations tackled in national or school-based student research projects do focus on real human problems and, as such, and because girls can work cooperatively at their own pace, are much more appreciated by girls. They also represent a better model for science to the boys.

Science and technology

The nature of the relationship between science and technology as taught in school is still being debated. At the one extreme, science is seen as a pure study of scientific knowledge and theories, and science practical work as the forming and testing of hypotheses. But such an approach forces science to become a limited, cerebral activity far removed from the activities with which most professional scientists, let alone engineers, occupy themselves. Their science is often much more applied, dealing with topics involving real people in everyday contexts. Their practical investigations are more concerned with solving problems than with testing hypotheses. For many practising scientists the gap between science and technology is very small and variable – indeed, science and technology might better be thought of as the ends of a continuum rather than as two distinct entities. There are, of course, distinctions that can usefully be made about science and technology (Layton, 1993), but technology must not be allowed to claim a monopoly of the goal 'to fulfil a human purpose'. Though much technology in the past has developed empirically, with little explicit reference to scientific understanding, that is unlikely to be so in the future. Efficient technology needs to be science-based. The development of science relies on a utilisation of technology.

At the school level, however, the integration or harmonisation of science and technology is far from easy. This is due partly to the different sub-cultures and rivalries of science and technology teachers and partly to the formalisation of the curriculum which has, in England and Wales, defined and assessed technology as a subject quite distinct from science. In other countries physics has been associated more closely with technology, but this neglects the technological aspects in chemistry and biology. It turns technology into

engineering and omits the study of the technologies needed in food, health, medicine, pollution and the environment. Her Majesty's Inspectorate (HMI, 1985) said of technology:

> The essence of technology lies in the process of bringing about change or exercising control over the environment. This process is a particular form of problem solving: of designing in order to effect control. It is common to all technologies including those concerned with the provision of shelter, food, clothing, methods of maintaining health or communicating with others, and also with the so-called high technologies of electronics, biotechnology and fuel extraction and the alternative technologies . . .

and such problems can only be solved if science is brought together with technology. Though the problem-solving cycle in science (after APU) is not dissimilar to the design process in technology, the types of investigation that can be done in school science are necessarily more limited than the projects allowed in the larger time allocation of technology – especially under the constraints of the National Curriculum's first Attainment Target on Scientific Investigation (DES, 1991). The encouragement of extended investigations for the higher levels of Sc1 in the National Curriculum, and the projects incorporated in some of the A-level science courses (Nuffield physics and biology, and University of Cambridge Local Examinations Syndicate (UCLES) Modular Sciences), offer ways of overcoming this problem and are much welcomed.

It may be that pure science in schools is effective for a minority of potential scientists. But for the majority of students – for all future citizens and those going towards applied sciences such as medicine, geology or engineering – the setting of science into its broader technological context, dealing with real problems involving real people, is necessary to make it effective and satisfying.

The development of the technology curriculum in the National Curriculum for England and Wales has not been an easy one (Eggleston, 1992). It was an ill-defined subject, with many faces depending on the particular tradition of those teaching it

(Smithers and Robinson, 1992). The original version has not been warmly received, having been criticised as too detailed, too jargon-laden, and insufficiently related to the needs of industry and the engineering profession. One fundamental flaw in its design was the absence of any attempt to relate it to the previously developed science curriculum. We await (1993) with interest to see if the revised version will prove more satisfactory.

We will show that many of the aims and methods of science and technology are common, that the two are interdependent, and that science and technology can be brought together effectively both in the formal curriculum and, even more, through the extended projects and investigations done in extra-curricular science activities and in school–industry links.

Organisation and teaching issues

Science or sciences

One of the most significant changes to science teaching in England and Wales in recent years has been the move away from teaching biology, chemistry and physics as separate subjects, instead treating science as a single subject. Though the content of the curriculum has not changed dramatically, the implications of this organisational change have been great. The reasons for the change have been many and varied: a mixture of educational and administrative reasons and expediency. I have discussed these at length elsewhere (Woolnough, 1988). Suffice to say that there are both advantages and disadvantages in this transition. In the late 1980s only a minority (10 per cent) of students studied all three sciences up to the age of 16, in approximately 30 per cent of the curriculum time. Now all students are studying a balanced science course, the majority in approximately 20 per cent of curriculum time.

The advantages of this are strong: all students are being given a more balanced coverage of science, including not only biology, chemistry, and physics but also geology, astronomy and the nature of science; the common aspects of science

are stressed; topics are treated in a more multi-disciplinary way; and science departments are working together more closely, with experts in the different sciences providing course material and advice for those from other disciplines.

The disadvantages are the obverse of the advantages. Some students, particularly the more scientifically able who were previously doing three sciences separately, are spending less time on science; the distinctive aspects of the different sciences are reduced; topics are treated in less depth; and teachers are teaching outside their own disciplines more, and thus with less expertise and enthusiasm. Whereas it is not too difficult for any science teacher to teach all of the sciences together at the lower age range, not all teachers feel that they can do so, enthusiastically and expertly, for the older students, where more specialist knowledge is needed.

There are different ways of delivering balanced science, through integrated courses, coordinated courses, modular courses, or the three separate sciences. Schools will seek to choose the appropriate approach to utilise the particular strengths of their science teachers. Suffice to say that experience and research are showing that it is no trivial matter to expect every science teacher to teach all three sciences; that the sciences have important distinctions as well as certain common approaches; that many teachers find they can teach their own science more effectively and more enthusiastically than others, and that teaching the other sciences is more difficult, more time-consuming, and less satisfying; and that students with the potential to continue with the sciences professionally also react to the sciences, and the science teachers, differently: many students, for example, are influenced into physics by an inspired and enthusiastic physics teacher, others into biology by a good biologist. Chemists pass on their enthusiasms to future chemists. The pressures on science teachers are very high: some have found unreasonable the expectation to teach all of the sciences up to age 16. Innovation fatigue in science departments due to the introduction and assessment of balanced science in the National Curriculum is not uncommon. There is evidence that it is forcing out from the science programme of some schools just the very activities that are the most effective – the extra-curricular science activities, clubs, competitions and even individual pupil–teacher conversations. We will show that a judicious use of teachers, working within their own confidence and subject expertise, enables them to have the time and enthusiasm for extra-curricular activities, and also to teach science most effectively.

Styles of teaching

It would be useful to know what is the most effective way of teaching science. Should we concentrate on practical work, and if so, should the teacher impose a meaningful structure or encourage the students to take more responsibility for their planning? Should we concentrate on written work, on dictated notes or on free student writing? Are written exercises better than extended projects? Is a lecture demonstration more effective than a group discussion? Is mixed-ability teaching more effective than streamed or setted classes? Much research has been done in the past to try and find which is the most effective method, and most has been inconclusive.

Effectiveness depends on the aim of the teaching, on the individual student, and on the teacher. Taking aims first, if the aim is to develop practical skills, then practical exercises will be more effective than watching videos. If the aim is to develop students' powers of self-expression, then opportunities for free writing will be more effective than dictated notes. If we want students to understand some difficult theory then teacher explanation will be more effective than an open-ended practical. If we want students to be able to use science in solving mathematical problems, then we will need to give them practice. The most important factor is to be clear about the aim for a particular lesson before deciding on the most appropriate style of teaching.

However, different students prefer and respond differently to different teaching styles. Some prefer to be given freedom and responsibility to plan their own work, others prefer the security of a more structured approach. This will be affected by their psychological make-up, their cognitive development and their emotional state. For any given class a mixture of teaching styles will be needed, satisfying different students differentially with different types of activity.

Teachers, too, have different personalities and different preferred teaching styles, and that affects their effectiveness. Most styles, delivered well, will be effective. (Most styles, delivered badly, will be ineffective!) One teacher's teaching style will not necessarily match anothers'; all need to find out where their strengths and weaknesses lie and develop their teaching styles accordingly.

Underlying the quest for the most effective teaching style is the question of how children learn. This is not the place to discuss that at length; since the 1970s, psychologists have made considerable progress in attempting to understand the way children reorder their cognitive structures, but there is still much that we do not know. The work of Piaget has given us insights into the different types of knowledge and when students can best accommodate it (Shayer and Adey, 1981). The work of Ausubel (1968) has shown us the importance of building on students' prior knowledge and the constructivists have helped us in uncovering that prior knowledge and taking it further (Driver and Bell, 1985). The writings of Polanyi (1958) remind us not only that explicit knowledge is important, but also that students' tacit knowledge is vital for much problem-solving, especially in science. As he put it: 'We know more than we can say'. Linguistic psychologists (Barnes, 1972; Sutton, 1991) have shown us the importance of language in learning: that language is more than a vehicle for communication of what is known; it is also a means whereby we sort out our own thinking, by 'talking ourselves, or writing ourselves, into understanding'.

Underlying most of these insights is the principle of active learning. For students effectively to learn, appreciate, personalise and remember something they must take an active part in learning it. It will not be acquired by passive learning.

Meaningful learning is not acquired by treating the mind like an empty pot to be filled; it is more akin to going on a journey.

But beyond, and underlying, the cognitive dimension of learning is the affective. The importance of the students' commitment, determination, enthusiasm and self-confidence cannot be over-estimated. No learning will take place unless the student is willing and committed. No potential will be realised unless the student responds to a challenge. It is in this area that we, as science teachers and educators have most to learn. No matter how good our curriculum, how cognitively correct our teaching methods, unless we are able to motivate our students to enthuse about their science and make a commitment in it, we will have given them little of lasting importance.

We will show that it is possible to build on a variety of appropriate in-class teaching styles, which equip students with knowledge of and enthusiasm for science, and that through individual projects and research investigations students can consolidate and develop that enthusiasm into commitment and personal achievement.

Doing or thinking

Many of the organisational issues which have dominated science teaching in England and Wales have been based on the principle that *learning* science is about *doing* science, and that the best way to learn science is by doing practical activities in science. Teachers have almost felt guilty if they have not had their classes doing practical work in any particular lesson. The old adage 'I hear and I forget, I see and I remember, I do and I understand' is deeply ingrained in school science in England and Wales. Despite various analyses of the dangers of doing too much, or inappropriate types of, practical work (Wellington, 1981; Wool-nough, 1991b), there is still much unfocused practical work which is far from effective. For many, still, the doing does little to increase the cognitive understanding – in Ros Driver's adaptation of the above adage: '. . . I do and I get even more confused'. With all the distracting clutter of

irrelevant detail in practical experiments, without taking note of the selective effect of students' prior knowledge, and with no clear purpose to address to the practical activity, it is not surprising that students find it difficult to discover the profundities of scientific theories from their practical work. We should not forget that it took even scientists of the calibre of Dalton, Curie, Mendel, Lister, Newton and Faraday a long time to discover what we expect our students to in a couple of periods!

The interaction between theory and practical work is a sensitive, iterative process. Students see in the practical what they have been prepared to see, either through their own preconceptions or by focused prepared questions. They then modify their prior knowledge, if they must, and need to develop it through teacher-directed discussion and student application and practice. This refinement and improvement of theoretical understanding will need to be done in class and away from the laboratory equipment, where students are obliged to think rather than do. The refined insights, and sound theoretical understanding, can then be taken into further practical investigations where they will be consolidated and used in solving problems. Practical activity thus has two functions: to consolidate theoretical understanding and to develop competence at, and confidence in, practical, scientific problem-solving. A tension arises when these two aims are not distinguished and it is hoped that the same practical activity will produce the attainment of both, in which case neither is achieved satisfactorily.

But school science is more than doing science and understanding the facts and theories of science. It is also about understanding the way scientists think, and this centres on the language, metaphors and models that they use. Clive Sutton (1992) discusses these issues and argues

> for a reinstatement of words as instruments of understanding, that they [words] cannot be taken for granted while we busy ourselves in the organis-ation of practical work.

Enabling our students to become scientists, even 'scientists for a day', challenges us to organise their learning so that they both act and, increasingly, think like scientists. We hope to show that some extra-curricular science activities, such as visits, lecture demonstrations and joint school–industry projects will enable students to learn something of the way scientists think as they work with and listen to real scientists and engineers in action.

CHAPTER 2

Insights from schools

In the attempt to find answers to some of the questions raised in the previous chapter, in particular to find out more about the factors that influence students' attitudes to science, I conducted some research into the views of students and their teachers, and the factors that affected their response to different types of school science. The research focused primarily on those students who were continuing with science into and beyond sixth-form study. One aim of school science teaching (though certainly not the only or even, arguably, the most important aim) is to encourage and prepare students for careers in science and engineering. I was primarily concerned to find those factors which influenced students positively towards such careers. Fortunately, I found that most of those factors are the same ones which positively influence the vast majority of students, whether or not they are to continue with science into science based careers. Thus these research findings relate to effective science teaching in the wider sense – effective in encouraging many students to enjoy and continue with science beyond school and college, but also effective in providing all students with a satisfying and stimulating experience which will enable them better to appreciate and function in the world in which they live. The full results of this research, the FASSIPES (Factors Affecting Schools' Success in Producing Engineers and Scientists) project, are published elsewhere (Woolnough 1991a; 1993; 1994). This chapter will give a summary of that work as it relates to our discussion, for there are some important findings

which indicate the way science teaching should develop in schools and colleges if it is to become more effective.

The research structure

The research sought to address two central questions. First, what are the factors which influence students towards careers in science and engineering? Second, what can schools do to make science teaching more effective? Some factors which influence students' choices are clearly outside the control of the school, factors such as job opportunities and family background. But are there influential factors which teachers *can* control? Thankfully the answer to this question is yes, schools do make a difference, and many teachers have developed ways of encouraging students in science. Despite horror stories in the media to the contrary and despite the restrictions and burdens imposed upon teachers by the National Curriculum and assessment, science teaching is alive and kicking in many schools and many students find their experience of science in schools both stimulating and fulfilling.

The research questions were addressed in two phases. First, heads of science in schools throughout England were sent a questionnaire asking them about the way science was taught in their schools, the numbers of students studying and hoping to continue with science into higher education, and the factors that they thought influenced young people towards or away from the sciences and engineering. It was hoped that there might be some correlations between the way schools taught science and their success in encouraging students to become scientists and engineers. Subsequently, in the second phase which built upon the responses of heads of science, case studies were carried out of the science departments of a dozen schools and colleges which were successful in teaching science. Staff and sixth formers were interviewed. Students were asked about themselves and their career aspirations, about the factors affecting their own decisions, about how they responded to different types of science teaching, and about the factors that affected their career choices. It was hoped that the students and teachers might be able to give fuller insights into the causal relationships and underlying statistical correlations which affected their choices.

In all, 132 heads of science completed the questionnaire, 12 schools were visited, 108 sixth-formers and 84 staff were interviewed individually, and questionnaires were returned from 1,210 upper sixth-form students. The findings, insights and conclusions reported below are based on an analysis of the data received and a synthesis of the various perceptions of teachers, students and researchers.

Initial findings

In phase one, the heads of science were asked about the type of science they taught, and about their use of different approaches. Did they encourage their students to plan their own experiments or to follow instructions on prescribed worksheets? Did they have their students write up their notes in their own words or use a common form of words? Did they use extended projects? Did they make use of science clubs or competitions? Did local engineers or parents get involved with school science? A set of 19 statements described such different approaches,

and teachers were asked to rate their use of them on a five-point scale.

When the responses were analysed using factor analysis it was found that three strong and sensible groupings were produced which described three specific types of approach to science teaching. The three were titled 'student-centred', 'teacher-centred' and 'extra-curricular activities'.

The student-centred schools encouraged students to plan their own experiments and activities, encouraged student self-assessment, used extended projects in years 12 and 13, and encouraged students to write their notes in their own words.

The teacher-centred schools encouraged students to write their notes in a formal style, used well-structured and teacher-directed science lessons, gave experimental instructions to students on worksheets and assessed all students' work against teachers' mark schemes.

The third group of schools, the extra-curricular activities schools, had science clubs and science and technology competitions for their students, involved parents and local engineers in the work of the science departments and, at times, suspended the normal timetable over an extended period of time for projects.

A simple measure of 'success' was defined for the schools, measuring the proportion of those students achieving at least five GCSEs who went on to higher education to study one of the physical sciences or engineering. It was thought that there might be some connection between the way schools approached science teaching and the number who continued with it into higher education, as indicated by this measure of 'success'. In fact the results showed that there was no preference between the two styles of teaching: both student-centred schools and teacher-centred schools were similarly 'successful', though there did seem to be some indication that pure scientists (those going on to study physics or chemistry in higher education) correlated more with the student-centred schools, while the applied scientists (those going on to engineering courses) correlated more with teacher-centred schools.

There was, however, one very strong and

significant correlation. The extra-curricular activity schools were the ones which had a relatively high degree of 'success'. Those schools which encouraged extra-curricular activities and student science projects, through clubs, competitions, projects and school–industry links, were the ones which sent a large proportion of their students on to higher education to continue with their sciences or engineering.

Of course, statistical correlations are not the same as causal relationships. A correlation between two factors does not prove that one causes the other. However, there is evidence from the second phase of the research that for many students there is also a causal connection here. Many students say that the things which really 'switched them on' to science were just these extra-curricular science activities and the student research projects which challenged them and fired their imagination. But we will return to that later.

There was one other factor which showed a very high correlation with the schools' 'success' in encouraging students to continue with science or engineering into higher education, and that related to the way schools organise science teaching. Schools which strongly encouraged students to take three separate sciences in years 10 and 11 (rather than teaching integrated science) correlated strongly with 'success' (significant at the 1% level). Again, we need to be careful in interpreting this correlation: it does not imply that one causes the other. In subsequent aspects of this research we have, however, been able to 'unpick' the causal links here. The link is through the key importance of expert teachers teaching subjects in which they are confident and enthusiastic. In practice, this happens more in schools where well-qualified specialist teachers have been able to teach their own specialism in years 10 and 11 than in schools where the science teachers, perhaps less expert in one of the separate disciplines, have had to cover the whole range of the sciences in an integrated science course. Teachers obliged to teach science in an area outside their own competence are unlikely to do so with enthusiasm and conviction and are thus unlikely to be effective.

It should also be noted that teaching the sciences as separate sciences may be more effective in encouraging the more able students to continue with sciences but may be less effective in providing an appropriate science course for the majority of students for whom continuing with science studies even beyond the age of 16 may not be appropriate. Our research did not consider this possibility directly, though there is evidence that where the separate science courses have continued to be dominated by the traditional top-down approach (that is, one determined by the goal of preparing some students for higher education courses), it has maintained an abstract, academic approach which is neither appreciated by the majority of students nor relevant to their lives as educated citizens. There is a perceived conflict of interests here between those being prepared for academic courses in higher education and those wanting science for life. We will show, however, that this conflict exists more in the perception of university scientists than in reality, and that there need not be a conflict between relevance and rigour, interest and academic grounding, if students are challenged to respond to the best of their individual ability and given the freedom to develop in the way appropriate to them.

The heads of science were also asked what they thought were the factors which encouraged and those which discouraged students in following up sciences or engineering into higher education, and these, too, were illuminating (see Figures 2.1 and 2.2). They believed that the encouraging factors centred on good science teaching and encouragement from science teachers; on the nature of science itself (its practical, challenging and problem-solving nature); on extra-curricular science activities (speakers, visits, clubs and competitions), as well as on such external factors as home background and student ability. They believed that the most discouraging factors were centred on the difficulty of the physical sciences in schools, especially in the A-level courses, and on the perceived low salaries and status of jobs in science and engineering.

Subsequent work in the second phase of the

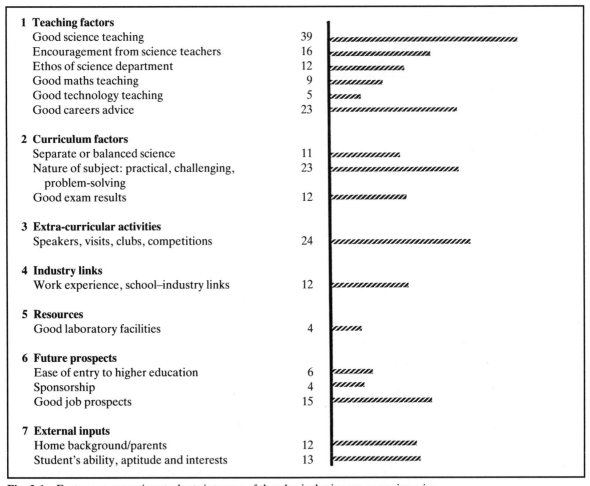

Fig. 2.1 Factors encouraging students into one of the physical sciences or engineering

The number after each factor records the number of teachers who mentioned that factor

research has shown, however, that although these generalisations are true for many students, some very able students are put off science not because it is too difficult but because it is too easy and not intellectually challenging enough. In addition, the perception of careers in science and engineering as having low status and salaries is stronger in non-scientists than in scientists.

So even in this first phase of the research there are indications that effective science teaching has much to do with extra-curricular activities in science, competitions and research, as well as with stimulation and personal encouragement from the science teachers themselves. These messages will be strengthened by subsequent evidence.

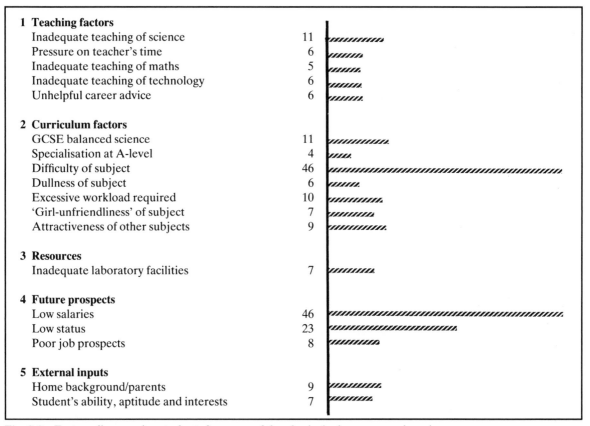

1 Teaching factors	
Inadequate teaching of science	11
Pressure on teacher's time	6
Inadequate teaching of maths	5
Inadequate teaching of technology	6
Unhelpful career advice	6
2 Curriculum factors	
GCSE balanced science	11
Specialisation at A-level	4
Difficulty of subject	46
Dullness of subject	6
Excessive workload required	10
'Girl-unfriendliness' of subject	7
Attractiveness of other subjects	9
3 Resources	
Inadequate laboratory facilities	7
4 Future prospects	
Low salaries	46
Low status	23
Poor job prospects	8
5 External inputs	
Home background/parents	9
Student's ability, aptitude and interests	7

Fig. 2.2 Factors discouraging students from one of the physical sciences or engineering

The number after each factor records the number of teachers who mentioned that factor

Students' opinions on science activities

One of the most important probes in the second phase of the research was the questionnaire directed at year 13 students, half-way through their final year in a range of schools and colleges. At this stage they could look back on their whole school experience of science, and have a good idea as to what they were going to do after leaving school at the age of 18 or 19.

They were asked about their achievements and expectations for the future, about how they reacted to the types of science they had experienced through their years at school and about the factors that had encouraged them into or discouraged them from science and scientific careers. Their responses showed a surprising consistency in some respects and significant differences in others. The responses were analysed for different types of school and college (which showed remarkable similarities), for different sexes (at this stage, the boys and girls reacted in a very similar way to the different types of science experience they had had and in terms of those factors which had been encouraging or discouraging), and for different categories of student (in terms of the subjects to be studied at higher education).

Table 2.1 summarises the students' responses

Table 2.1   Mean student response to student activities in school science

The students were asked to rate on a five-point Likert scale how they reacted to the following numbered statements about different types of activities in school science, from 'strongly agree' (5) to 'strongly disagree' (1). The higher the score the more the students agreed with the statement. A score of 3 corresponds to the midpoint of the scale. Those responses marked p had a distinct polarised response, with two distinct peaks. Any significant difference between scientists and non-scientists is marked * (<5%), ** (<1%), or *** (highly significant, <0.1%).

	All	Boys	Girls	Scientist	Non-scientist	
	1180	669	511	458	722	
1 I found the opportunity to plan my own experiments very satisfying.	3.2p	3.2	3.1	3.4	3.1	***
2 I felt happiest when clear instructions were given to follow when doing practical experiments.	3.9p	3.8	4.1	3.9	3.9	
3 School science should be about learning scientific facts and theories.	3.0p	3.0	2.9	3.2	2.9	***
4 School science should be about learning to do science through scientific investigations.	3.9	3.9	3.9	3.9	3.9	
5 Standard experiments, written up correctly, give confidence to continue with science.	3.5p	3.4	3.6	3.5	3.5	
6 Extended practical projects showed me what science was like and got me interested in it.	3.0	2.9	3.0	3.2	2.9	***
7 The best notes are short and concise.	4.0p	4.0	3.9	4.1	3.9	*
8 I feel I need to write quite a lot to really express myself satisfactorily.	2.9p	2.8	3.1	2.8	3.0	
9 I feel most confident when the science lessons were well structured and teacher-directed.	3.8	3.8	3.9	3.9	3.8	**
10 I valued the opportunity when the teacher let us plan our own activities in lessons.	3.1	3.2	3.1	3.2	3.1	
11 Student work should be marked objectively by the teacher.	4.1	4.1	4.2	4.2	4.1	
12 The most effective form of assessment is self-assessment by the student.	2.5	2.5	2.5	2.5	2.5	
13 The times when the school suspends its normal timetable for extended projects are not very useful.	2.9	2.9	2.9	2.8	3.0	
14 Involvement in science clubs is an unhelpful distraction from the learning of real science.	2.7	2.7	2.6	2.6	2.7	
15 Parents should not be involved in the work of the school science department.	2.8	3.0	2.6	2.8	2.8	
16 Involvement in science and technology competitions is great fun and useful.	3.2	3.2	3.3	3.4	3.1	**
17 Local engineers can bring a stimulating dimension into science lessons.	3.3	3.2	3.4	3.3	3.3	
18 Work experience in science-based industry turns people off jobs in science or engineering.	2.6	2.7	2.5	2.5	2.7	***
'Student-centred'	19.8	19.7	19.9	20.6	19.3	*
'Teacher-centred'	7.3	7.1	7.7	7.3	7.4	

Table 2.2   Group factors relating to student activity in school science

---

'Student-centred' = + Plan Exp (1) + LearnDo (4) + ExtProj (6) + PlanAct (10) + Compet (16) + LocalEng (17)

'Teacher-centred' = + WrkSht (2) − LearnDo (4) + StrExp (5) + Struct (9)

Numbers refer to statements in Table 2.1

for each of the statements about how they reacted to different types of scientific activities they had experienced in school. It shows the average scores for all students, scores for boys and girls taken separately, and scores for those going on with their sciences, or engineering, and those who were not going to be scientists. It also notes the relative scoring of those students who preferred 'student-centred' teaching and those who preferred 'teacher-centred' teaching. These groups of students were revealed by a factor analysis of the data and showed preference for the characteristics as described in Table 2.2.

The students were asked about the types of scientific activity they had experienced in schools, corresponding to those that the heads of science had been asked about previously. They were given statements about each of them and asked whether they agreed or disagreed with the statements, on a five-point scale. The overall picture was a fairly conservative one, with the students preferring to have well-structured, teacher-directed lessons than ones in which they were expected to plan for themselves and take their own initiative. Students agreed most strongly with those statements which said they felt happiest having clear instructions to follow when doing practical experiments, they felt most confident when the science lessons were well structured and teacher-directed, and felt that student work should be marked objectively by the teacher. They also felt strongly that the best notes are short and concise and that school science should be about learning to do science through scientific investigations.

Surprisingly, perhaps, there was no significant difference between the boys and girls at this stage

as they looked back over their science experience. By this time, however, only the more able and persistent students were still in school or college, so that the sample did not accurately represent the full range of boys and girls studying science in the earlier stage of secondary schooling. Students were older and more mature and thus less liable to be 'gender-polarised' by peer-group pressure.

One of the most important factors which should never be overlooked in reporting research focusing on means and average trends, is the great dispersion of response around these means. Different students react quite differently to the same teaching strategy, and even the same student may react differently to the same strategy presented in a different context and on a different day.

Different groups of students differed in their reactions to these different approaches, too. Potential physicists, in particular, seemed to prefer more independence of thought and action. Compared with non-scientists, scientists generally found the opportunity to plan their own experiments very satisfying, agreed that extended practical projects showed them what science was like and excited their interest in it, and had found involvement in science and technology competitions to be great fun and useful. A factor analysis of the students' responses produced a similar grouping to that of the schools in the first phase of the research, a student-centred group and a teacher-centred group. The student-centred group liked planning their experiments and activities, thought that school science should be about learning to do through scientific investigations, found that extended projects showed them what science was like and excited their interest in it, had found

Table 2.3   Mean student response to encouraging or discouraging influences

The students were asked to rate on a five-point Likert scale how they had been influenced by the following numbered items in their choice of career and higher education in one of the physical sciences or engineering, from 'very positive' (5) to 'very negative' (1). The higher the score the more the students agreed with the statement. A score of 3 corresponds to the midpoint of the scale. Any significant difference between boys and girls, and between scientists and non-scientists is marked * (<5%), ** (<1%), or *** (highly significant, <0.1%).

	All	Boys	Girls	Scientist	Non-scientist	
	1180	669	511	458	722	
19 The quality of the teaching in the science department	3.3	3.3	3.3	3.7	3.1	***
20 The personal encouragement given by science teachers	3.3	3.3	3.3	3.7	3.1	***
21 Supportive maths teaching in the school	3.4	3.4	3.3	3.6	3.2	***
22 Supportive technology teaching in the school	3.0	3.0	2.9	3.1	2.9	**
23 Advice from careers staff	3.0	3.1	2.9	3.2	3.0	***
24 The practical nature of the science lessons	3.5	3.5	3.5	3.8	3.4	***
25 The intellectual satisfaction of doing science	3.4	3.4	3.3	4.0	2.9	***
26 The amount of involvement with human issues	3.2	3.0	3.4***	3.2	3.2	
27 The amount of self-expression allowed in science lessons	2.9	2.9	2.9	3.2	2.7	***
28 The tradition of good exam results in science	3.2	3.3	3.1	3.5	3.0	***
29 Outside speakers and visits to science firms	2.9	2.9	2.8	3.0	2.8	***
30 Local engineers coming into the school	2.9	2.9	2.8	3.0	2.8	***
31 Work experience in local companies	3.0	3.0	3.0	3.1	3.0	**
32 Involvement in science clubs (photographic, radio, etc.)	3.0	3.0	3.0	3.0	3.0	
33 Involvement in science competitions (e.g., great egg races)	3.0	3.0	3.0	3.1	3.0	*
34 The level of difficulty of the sciences at school	3.1	3.1	3.0	3.4	2.9	***
35 The amount of work required for school sciences	3.1	3.1	3.0	3.2	3.0	***
36 The ease of entry to HE for science and engineering	3.0	3.1	2.9	3.2	2.9	***
37 The possibility of sponsorship in HE	3.1	3.1	3.0	3.2	3.0	***
38 The status of jobs in science and engineering	3.2	3.2	3.2	3.4	3.0	***
39 The likely salaries in science and engineering jobs	3.2	3.2	3.2	3.4	3.1	***
40 The likely job satisfaction in science and engineering	3.3	3.4	3.2	3.9	2.9	***
41 The sophisticated technology used in military weapons	2.8	3.0	2.6***	2.9	2.7	*
42 The situation in local science-based industry	2.9	2.9	2.9	3.0	2.8	**
43 Experience of your family in science-based industry	2.9	2.9	2.9	3.1	2.8	***
44 Scientific hobbies and fiddling with gadgets at home	3.3	3.4	3.0***	3.7	3.0	***
'Extra-curricular activities'	15.1	15.1	15.1	15.5	14.9	***
'In-class activities'	22.8	22.9	22.8	25.1	21.4	***
'Career aspirations'	9.6	9.8	9.5	10.7	8.9	***
'External factors'	11.9	12.2	11.4	12.7	11.3	***
'Difficulty of subject'	6.1	6.2	6.1	6.6	5.8	***
'Higher education incentive'	6.1	6.2	5.9	6.5	5.8	***

involvement in science and technology competitions to be great fun and useful, and had appreciated the involvement of local engineers. Future scientists were slightly more student-centred than the non-scientists, but such an approach was widely appreciated by all students, where they had experienced it.

At first there seems to be some contradiction

Table 2.4 Group factors relating to encouraging and discouraging influences

'Extra-curricular activities' = +SpAndVis (29) + LocEng (30) + WkExp (31) + ScCl (32) + ScCompet (33)

'In-class activities' = + QualTec (19) + TeacEnc (20) + PracNat (24) + IntSat (25) + HumanIs (26)
+ SelfExp (27) + GdExams (28)

'Career aspirations' = + Status (38) + Salary (39) + JbSatisn (40)

'External factors' = + Weapons (41) + LocalSBI (42) + FamExp (43) + Hobbies (44)

'Difficulty of subject' = + DiffOfSc (34) + WrkInSc (35)

'Higher education incentive' = + HEEntry (36) + Spons (37)

Numbers refer to items in Table 2.3

here. The students prefer to have their science lessons well structured and teacher-directed but they also respond to some opportunity to do investigations, take part in competitions and plan their own experiments. But this apparent dichotomy is, I believe, at the centre of effective science teaching. Such teaching should not be either entirely teacher-directed or entirely student-centred, but should combine the two. Students do need the structure to build confidence in their scientific knowledge and techniques, but they also need the opportunity to test themselves out, to be given some personal challenge, through investigations, projects and competitions.

Students' opinions on influencing factors

The students were also asked about factors that would have influenced their career choices, factors that the teachers had suggested were important in the first phase of the study, and asked whether these had been encouraging or discouraging factors for them. As might have been expected, there was considerable variation in their responses – some were influenced by some factors, others by others; some would have found a certain factor encouraging, others would have found the same factor discouraging. However, factor analysis revealed six, quite distinct, groupings of students who seemed to be influenced by quite different factors. Table 2.3 summarises the overall student

responses to the influencing factors, showing the average scores for all the students, for the boys and girls taken separately, for those going on with their science/engineering, and those not becoming scientists. It also notes the different scores for the six groups of factors which were identified and described by the factor analysis. These groups showed strong and meaningful groupings of the different factors as listed in Table 2.4.

Overall, the most encouraging factors were the quality of the science teaching and the encouragement of the science teacher, good maths teaching and the practical nature of science, the intellectual satisfaction in science lessons, the likely job satisfaction and scientific hobbies and fiddling with gadgets at home!

Most of the factors suggested had, unsurprisingly, a more positive response from the potential scientists than from those who had rejected science. The biggest difference between those who were to continue with science and those who were not related to the intellectual satisfaction of doing science: some found it great and were continuing with it, others didn't and weren't! We found other evidence, too, that many potential scientists did not want the science content watered down and made easier; on the contrary, they responded very positively to the intellectual stimulation. Future scientists also differed considerably from non-scientists in the perceived job satisfaction in science (self-evidently!) and in the amount of time they had spent in scientific hobbies and fiddling with gadgets at home. Domestic background is very influential here, with significant differences, between the sexes.

There were, however, three factors which had an equal response from scientist and non-scientist alike: the amount of involvement with human issues; involvement in science clubs; and involvement in science competitions. These factors, though important as a stimulus for future scientists, clearly have a more universal appeal, too.

The fact that there were six distinct groupings of influencing factors revealed by the factor analysis (described in Table 2.4) is interesting in that each grouping contains different sets of factors. This revelation is important because it shows that there is not one single factor, or group of factors, which is influential on all students; some students are influenced towards scientific careers by one group of factors, other students by quite different factors.

Furthermore, the different groupings of influencing factors affected different types of students differently. The type of student was defined according to what the students were (or were not) going to study in higher education. Ten student types were defined, and these are listed in Table

Table 2.5   Grouping of student types

The students were grouped according to their higher education (HE) intentions as follows:
Student type 1   = going to do physics (or astronomy) at HE
Student type 2   = going to do chemistry (or biochemistry or natural sciences) at HE
Student type 3   = going to do computer sciences at HE
Student type 4   = going to do (any type of) engineering at HE
Student type 5   = doing physical sciences but not in groups 1–4 above
Student type 6   = doing biological sciences but not in groups 1–4 above
Student type 7   = mixed arts and sciences, not more than one of maths, physics and chemistry
Student type 8   = arts students who had adequate GCSEs to have done the physical sciences
Student type 9   = arts students who did not have adequate GCSEs to have done physical sciences
Student type 10 = able arts students*

* As the arts students were on average less able than the physical scientists, a separate group (a subgroup of student type 8) was selected which was comparable in ability to the physics group, student type 1. Student type 10 was defined as having GCSE scores greater than or equal to 14 (seven grade Bs) and scores in maths, physics and chemistry greater than or equal to 6 (three Bs).

Table 2.6   Mean student response to encouraging and discouraging influences by student type

	Student type									
	1 Phys	2 Chem	3 CmSc	4 Eng	5 PhSc	6 BSc	7 Mixed	8 CDS	9 CNDS	10 Art+
'Extra-curricular activities'	15.4	15.8	15.0	16.1	15.3	15.2	15.4	14.8	15.1	14.6
'In-class activities'	25.7	26.4	23.8	25.0	24.2	25.6	23.0	21.4	21.2	21.2
'Career aspirations'	10.5	11.2	10.6	11.5	10.1	10.7	9.1	9.0	9.2	8.4
'External factors'	13.4	12.5	12.9	13.5	12.2	12.4	11.5	11.4	11.3	11.1
'Difficulty of subject'	7.0	7.3	6.4	6.6	6.7	6.3	6.0	5.9	5.8	5.8
'Higher education incentive'	6.5	6.6	6.4	7.1	6.2	6.2	5.7	5.8	5.8	6.1

2.5. Table 2.6 shows how the different types of student responded to the different groupings of influencing factors. Of the six groupings of influencing factors, the two strongest groupings, and the ones with the most interesting implications, are the extra-curricular activities group and the in-class activities group. The former are most affected by outside speakers and visits by local engineers and work experience, and by involvement with science clubs and competitions. The latter are most influenced by the quality of their science teachers' teaching and personal encouragement, by the practical nature of school science, its intellectual satisfaction, its involvement with human issues, the amount of self-expression allowed in science lessons and the tradition of good exam results in school. It is interesting to note that those going on to study engineering (student type 4) are most strongly influenced by the former, the extra-curricular activities, while the pure scientists, those going on to study physics, chemistry or biology (student types 1, 2 and 6) are most strongly influenced by the in-school activities.

So again the effectiveness of two types of science activity is evident. The in-class activity type: good, expert, sympathetic teaching of the school science curriculum in a humane, practical, intellectually satisfying and personally fulfilling way, which through role model influence encourages students to follow their science teacher into the pure sciences. And the extra-curricular type, encouraging creative problem-solving and links with the world of work; which seems to be particularly influential for potential engineers.

There were four other groupings which differentially influenced the career choice of students. One related to career aspirations, focusing on the likely status, salary and job satisfaction of scientific employment. A second related to external factors, the state of the local science-based industry and the experience of the student's family in it, the student's family background encouraging scientific hobbies and an awareness of uses of science in sophisticated technology. A third group of factors related to the difficulty of the subject, with the scientists finding the subject easy and the workload light, in contrast to the non-scientists who found science both difficult and hard work. A fourth and final group influenced by HE incentives, valuing the relative ease of access to higher education and the sponsorship available in their subjects. It was found that the engineers (student type 4) were most influenced by career aspirations, external factors, and HE incentives. It was the physicists and chemists (student types 1 and 2) who rated the difficulty of subject most positively, finding their science relatively easy and the workload more acceptable.

What the students say

What the students said in interview reinforced what had been indicated in the questionnaire survey: that they believed effective science teaching depended on the teachers and on the encouragement they received from them, on the way the subjects were taught and on the extra-curricular activities in science.

Do the teachers of science make a difference? Comments by two girls show the difference that teachers can make:

> I was put off sciences wholly by dull, unenthusiastic teachers.

> Having teachers who were interested and enthusiastic about what they were teaching helped to develop my love of science and learning about science.

Does the nature of the scientific activity at school make a difference? Again, comments from two male students, one who went on to study engineering, the other who could have done but did not, illustrate the difference and the perceived need to be personally challenged:

> There is too much theory and it is not really integrating the world with the subject. I'd prefer something where I could find out more for myself, I prefer to think things through for myself, prefer research to being spoon-fed.

Science lessons for GCSE were dull and uninspiring, allowing no use of imagination and creativity. Originality of thought was impossible, making it a dull process of studying existing views and pre-arranged or meaningless experiments.

Do extra-curricular activities in science make a difference? Not all students had experienced extra-curricular activities in their schools, but those who had, found, almost with exception, that they had been highly memorable and influential. Two boys, the first going on to study chemistry and the second engineering, illustrate the point:

I won the great egg race!

I went to a Bath University open day. There was a lecture on engineering after which I basically made up my mind that I wanted to go into aeronautics.

Do school–industry links make a difference? Many students had had some experience of this aspect, but there was a considerable variation in what they did and the effect it had. Well-planned initiatives were very impressive, but some industrial links, conveying the less stimulating side of industrial work, could be counter-productive. The following comments, both from girls, illustrate this;

Taking part in the Engineering Education Scheme convinced me to undertake an HE course in engineering, and to pursue a course in it.

For many years I wanted to work in engineering but after taking part in work experience in this industry I changed my mind completely.

Do the home background of students, and their ability and personality, make a difference? The main focus of our investigations was the effect of the school science department, but we were frequently reminded that there were certain highly significant external factors perceived as also being important. Home background, with one or both parents in science, medicine or engineering, clearly made a difference as the students grew up in an atmosphere where such careers were understood and respected, and where they were encouraged to take part in mechanical, biological or

'tinkering' hobbies. A very high proportion of those intending to go into higher education to study physics (47 per cent) or chemistry (45 per cent) had a relative working in science research or a science-based industry (53 per cent of those intending to study physics had at least one parent who had studied science or engineering beyond school). The perception that you need to be clever and of a certain type of personality was also strong. With the sciences, especially the physical sciences, it was often felt that you could either do it or you could not. Three quotes, from a boy going on to study engineering, a girl going on to study chemistry, and a boy intending to take physics, respectively, at university illustrate this;

I think it is probably my father's influence; he's an engineer and I've helped him around the house with a lot of things.

I think, if you do sciences, people think that you are cleverer than if you do other subjects. When people know that I'm doing physics, chemistry and maths, they go 'Wow!'

I did science because it was easy!

In the students' questionnaire the students were asked how they perceived their own personality against various personality traits. Figure 2.3 shows a summary of their responses, with the potential scientists and the non-scientists shown separately. Overall, these 18-year-olds had fairly similar perceptions of their own personality, with a few notable exceptions. The scientists saw themselves as cleverer, task-centred rather than people-centred, tough-minded rather than tender-minded, practical workers rather than abstract thinkers, more interested in ideas, more systematic, and communicating better in diagrams than in words. The non-scientists, those who had chosen not to study the sciences beyond the age of 16, were person-centred rather than task-centred, tender-minded rather than tough-minded, abstract thinkers rather than practical workers, very interested in people and creative, and better communicators in words than in diagrams.

Clearly the messages for science teachers are

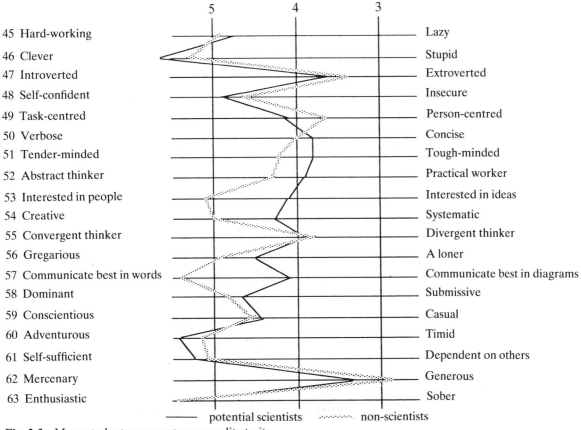

	5	4	3	
45 Hard-working				Lazy
46 Clever				Stupid
47 Introverted				Extroverted
48 Self-confident				Insecure
49 Task-centred				Person-centred
50 Verbose				Concise
51 Tender-minded				Tough-minded
52 Abstract thinker				Practical worker
53 Interested in people				Interested in ideas
54 Creative				Systematic
55 Convergent thinker				Divergent thinker
56 Gregarious				A loner
57 Communicate best in words				Communicate best in diagrams
58 Dominant				Submissive
59 Conscientious				Casual
60 Adventurous				Timid
61 Self-sufficient				Dependent on others
62 Mercenary				Generous
63 Enthusiastic				Sober

——— potential scientists   ·········· non-scientists

Fig. 2.3 Mean student response to personality traits

The students were asked to rate on a seven-point semantic differential scale how they perceived their own personality. An axis defined by two personality traits was drawn and the students invited to tick along the line. The higher the score, the more students perceived themselves nearer the first named characteristic. A score of 4 corresponds to the midpoint on the scale.

strong here. If we want science to be effective and attractive to those who currently reject it then we must take note of those who are primarily interested in people, and show that science is highly relevant to people's lives. We must also take note of the need in students to be creative and ensure that the science we teach gives encouragement to that.

Does the National Curriculum in science make a difference? It was surprising, perhaps, that few of the students actually mentioned the content of their science curriculum as being important. They spoke with enthusiasm about their teachers and the way they had been inspired and encouraged by them, they spoke of extra-curricular activities and projects and the way they had been challenged by those, but rarely did they speak of the specific curriculum. Perhaps the most depressing aspect of the whole research, however, was the cumulative effect that recent curriculum change was having on the energy and activities of science teachers. The increasing workload due to the assessment and administration of the National Curriculum was dampening the enthusiasm of many science teachers for their own teaching and reducing the time and energy that they had for extra-curricular

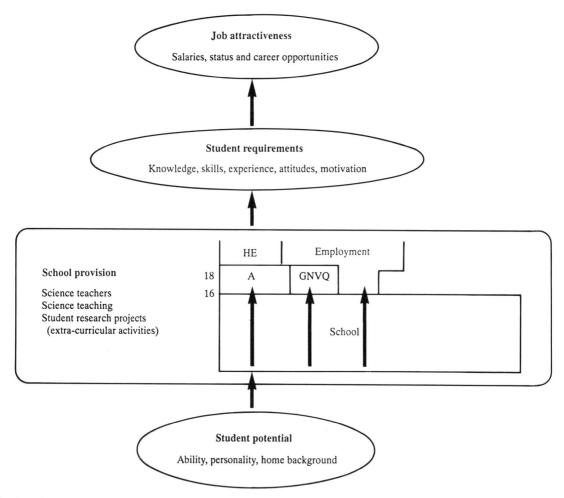

Fig. 2.4 Model for factors influencing students towards science or engineering

activities in science. Perhaps the saddest comment came from an able girl studying biology:

> We used to enter science competitions but we don't have time for that now.

We must hope that the present unproductive pressures that the National Curriculum is imposing on science teachers in England and Wales may soon be redressed and that teachers' energies may thus revert to the enthusiastic teaching of science and to the extra-curricular activities in science which are so effective.

A model for effective science teaching

Much of what has been described is summarised in Figure 2.4, which traces students as they come into a school, move through the science department and come out with a set of attainments which may or may not direct them into a career in science, medicine or engineering. Whether or not students do enter into such careers depends on a variety of factors, many of which are outside the control of the school. Students will be aware of the job attractiveness of careers in science and engineering, and this will reflect the importance that society and government give to such work. It will be measured in terms of salary, status and career opportunities. Currently in many countries, especially the UK, it cannot be said that the job attractiveness of careers in science and engineering is high.

Fortunately for science teachers, the worth of science teaching is not measured in terms of the number of students who take up careers in science. The goals of science teaching are shared to a large, though not total, extent for all students. The scientific knowledge and skills required for a scientifically literate citizen by the age of 16 are not vastly different from those required by a future scientist or engineer – though the differences are there and need to be addressed. More important still are the positive experiences of scientific activity over a wide range of contexts and the attitudes and motivation that students can acquire.

Attitudes of self-confidence, enjoyment of, enthusiasm for, commitment to and mature criticism of science and scientific activity, which are important for careers, are also of great importance to any adult life.

Children entering school will inevitably come with different potentials, different abilities, different personalities and different home backgrounds and these will significantly affect their reaction to school science. To some extent these can be built upon or compensated for by the school provision. As is beginning to become clear, the key factors here are not the fine detail of any science curriculum that may be laid down, but the interaction that lively science teachers will have with their students and type of science activity the students are involved in, both inside and outside the school science laboratory – especially those extra-curricular activities and research projects which stimulate, challenge and fulfil students' needs.

CHAPTER 3

So what makes the difference?

Successful science teaching

The main findings of the FASSIPES (Factors Affecting Schools' Success in Producing Engineers and Scientists) research described in the previous chapter are of fundamental importance. We now have a better understanding of what makes effective science teaching. In essence, it consists of two factors: good teaching in the classroom and (perhaps more surprisingly) effective use of extra-curricular activities. Good science teachers are: competent and enthusiastic in their subject knowledge, skilled and rigorous in their classroom management, and understanding and sympathetic to their students' needs.

Effective extra-curricular activities consist of student research projects and stimulus activities in science (which will be discussed more fully in Chapters 4 and 5 respectively). Through these activities students gain knowledge, understanding and appreciation of the sciences, confidence in and competence at doing science, and enjoyment, enthusiasm and commitment to the science appropriate for their own lives. Certainly, students' responses to school science will be highly influenced by their home background and the ability, attitudes and aptitudes that they bring with them into school. Some things can be done to encourage or supplement these. However, this book is concerned with what can be done in schools.

Good science teachers

Fundamental to all effective science teaching is the quality of science teachers, and their ability and opportunity to teach inspirational science. No matter how good the written curriculum is and the extent of back-up resources supplied, no matter how much incentive (through stick or carrot) is applied through the accountability of a national assessment scheme, no matter how much politicians exhort, unless good scientists enter schools and have the opportunity to share their enthusiasm for their subject with their students, science teaching will never become better than adequate. Indeed, it is possible that too much external pressure applied to the science teaching profession will make the job less attractive to good scientists and deter them from entering, or encourage them to leave, science teaching. There is evidence that 'innovation fatigue' has been occurring in recent years in the science departments of schools in England and Wales. The continuous change in science curriculum and assessment demands, with their accompanying excessive administrative workload, has been causing unreasonable pressure of work. One plea of this book is that we should simplify the procedures of science teaching so that we can concentrate on what is important, and allow good science teachers time to teach effectively.

Good science teachers are knowledgeable, competent and enthusiastic in their subject and in class management, and understanding and sympathetic to students and their needs. It is important, but not sufficient, to be an expert in the subject. It is important, but not sufficient, to be able to deliver

Box 3.1   What do students think a good science teacher is?

Quotes from the FASSIPES research described in Chapter 2.

What makes a good science teacher?

Having teachers who are interested and enthusiastic about what they were teaching helped to develop my love of science and learning about science.

A good teacher is someone who is actually interested in the subject and keen on it.

My physics teacher was a good teacher because she would try and relate physics to every day. The biology teacher was also quite good as she would take us out and try and relate things to the larger context.

In the third year I had a really good biology teacher. She was strict but she was nice. She really encouraged me and I started getting really good marks and I enjoyed it.

Good teachers . . . I find it easy to learn from them basically. It's just the feeling that you get from a teacher that he is good, I mean you don't fall asleep in a class if he's good.

Good teachers give you individual help if you're stuck on a particular topic, then their notes are good, you understand them.

A good teacher is someone who gets enough information across and is also friendly, with whom you can talk. Prefer people who can control the class.

A good teacher is able to explain complicated things to the required standard, and to go beyond the bounds of the syllabus.

Good teachers notice when you're confused and they help you. They know exactly what's going on and explain well.

Strict, someone who has time for you, who checks the homework, explains things well and has a sense of humour.

A good teacher's got to form a relationship with the pupils.

They are not only good at their subject, but they are anxious to make sure you understand. All of our teachers are very, very friendly. I prefer the teachers who take charge and make sure there is no mucking around, because I'm not attracted to fooling around so it's a bit annoying when somebody does.

Good teacher . . . knowledgeable, good with students and knows how to get them to do exams well.

My physics teacher was extremely supportive, very enthusiastic and he'd say, anytime come and see me. I had extra lunchtime sessions and he was always available.

Do students prefer teacher-directed or student-centred teaching? Some do, some don't!

I prefer to be spoon-fed information which I can learn.

One of the things I like is finding out why things happen, so I enjoy investigational work.

I prefer doing practicals as it is more active rather than when you have to sit and take it in. I certainly prefer to be told exactly what I'm doing, then I know I'm doing it alright.

I'd prefer something where I could find out more for myself, research more of it. I prefer to think through things for myself, prefer research than being told.

We've tried self-supported study, but it didn't sort of work with us. I prefer the teacher telling you, then you know that you are doing everything right, than doing something wrong and finding out about it afterwards.

One of my teachers in chemistry had given us the choice of working through a whole unit by ourselves, which we're doing at the moment, and we go to him if we have any problems when we're working, and I like that. It means that I can work towards something, I know where I have to go.

interesting and significant lessons. It is important, but not sufficient, to develop good relationships with students; to like, to respect and to understand them, and to help them to develop their potential. Good science teachers combine all three of these attributes.

Good science teachers like and enthuse about their own subject, and thus act as an inspiration and role model for many of their students. They keep up to date with their subject and are able to enrich their lessons with those fascinating extra-curricular nuggets which, while possibly slightly incomprehensible to some of the students, will show something of the excitement, the relevance and the continuing challenge that the subject can offer. They will want, and are able to find time, to talk with their students and to follow up their interests outside lesson time. Such conversations, not necessarily of long duration, can be of invaluable encouragement to students and affect their career destinations. They will find opportunities to bring their students into contact with other lively scientists, and possibly work with them in science projects (see Box 3.1).

Good science teachers are competent teachers and make their subject relevant, accessible and interesting for the student. Their classes are well ordered, their lessons and homework interesting and intellectually satisfying. They are rigorous and stimulating as well as accessible and enjoyable. They expect high standards from their students (higher perhaps than the students think they are capable of) but are at the same time sympathetic to the very real difficulties involved in learning and doing science. There has recently been much improvement in curriculum development to make the sciences less impersonal and more related to real people, less abstract and more relevant to the real world. Initiatives like *Science and Technology in Society* (SATIS) have shown science to be entirely relevant to students' lives in modern society (SATIS, 1987–92). Initiatives like *Salters Science* (Campbell *et al.*, 1990–) have shown that science can start with topics relevant and interesting to students and thence develop the concepts and theories which are important. Courses like

Active Science have shown that doing experimental work in science really is about conducting scientific investigations, rather than doing pointless practical exercises (Coles *et al.*, 1988–91). Science texts have changed radically, reflecting the message that science is no longer an arid, abstract, passive and impersonal subject but a lively, stimulating, active and human field of endeavour.

Good science teachers are also willing and able to build good relationships with their students: to be approachable, strict perhaps but not unfriendly, and to sympathise with the difficulties and the needs that students might have. They are patient with their lack of understanding of what may often be difficult concepts. They need, and have, a sense of humour! They will also appreciate the individual differences of students: their hopes and their fears; the times when they need support and encouragement and the times when they need challenge or correction; the times when they need the security of clear instructions and notes and the times when they can benefit by more freedom.

There is no space here to do justice to all the attributes that good science teachers bring to their lessons and their students. Undoubtedly, they are the key to effective science teaching, both in determining and delivering the curriculum, and in their relationships with their students in and out of class. For many scientists, engineers and medics, the most influential factor determining their career choice was an inspirational science teacher (see Box 3.2). For many who did not follow science into a career beyond school, their enjoyment and appreciation of science was also determined by their science teacher.

Extra-curricular science activities

But beyond the quality of the teachers and how they teach the science curriculum, there is a second group of factors which are found to be influential in effective science teaching which I have called, perhaps misleadingly, the extra-curricular activities in science. This is a broad category and not easily defined. Though most of the activities will be extra to the basic science curriculum, in some cases

Box 3.2   Top scientists inspired by enthusiastic teachers

Why did many of today's leading scientists become scientists? For many, the key factor was the inspirational effect that one of their science teachers had had upon them when they were at school. Tim Devlin and Hywel Williams (1992) researched the background of 1,600 people in Britain who had achieved fame in various professions and asked them what had influenced them into their careers. Among them was a group of scientists. The following quotes are taken from that book.

Sir Eric Ash; when at University College School, London, 'discovered enthusiastic scientists and mathematicians, two of whom were past normal retirement age but were carrying on during the war'.

Arnold Wolfendale, Astronomer Royal, 'came under the influence of A. N. Leaning, the physics master at Stretford Grammar School, who was very keen on practical hands-on experiments, had green fingers and made the experiments work – which predecessors had not always done'.

Sir Philip Randle, Professor of Clinical Biochemistry at Oxford, was 'inspired by an individual [science] teacher at King Edward VI Grammar School, who 'made it clear that he believed I had ability, and lifted me from indifference'.

John Thoday, Emeritus Professor of Genetics at Cambridge, said that 'the logical demonstration classes given by the physical science teacher, Mr Livesey, had a profound effect on my choice of genetics'.

Lord Lewis, Professor of Chemistry at Cambridge, 'was impressed and encouraged by a staff member who was dedicated to his chemistry' at Barrow Grammar School.

Denis Noble, Professor of Cardiovascular Physiology at Oxford, was taught at Emanuel School, London, by 'a superb team. The great strength of Mr Hurst, the chemistry master, was to be able to enthuse about the subject while going on with the demonstration. I remember his demonstration vividly. There was an element of showmanship in it but he made you realise that there was something very serious behind it . . . science was a serious business but it certainly was not boring. W. G. Garrard, the physics master, was quieter with less showmanship but he taught us rigour – that a scientist has to think rigorously – and if you did you were rewarded'.

Sir Richard Bayliss, a former physician to the Queen, 'had an absolutely inspired biology teacher called Peter Falk. His whole teaching was inspirational . . . he taught us never to accept anything as 100% certain because nothing ever was'.

Richard Lacey found that in the science department at Felstead 'there was an air of optimism with enthusiastic science teachers who did everything for you. The biology master Mr Sturdy explained things in the context of the real world . . . you would go on a field trip, come back and look at things under the microscope and then he would show their relevance to the real world'.

Ruth Jarrett, Head of the Leukemia Research Fund virus centre at Glasgow, enjoyed most the times 'when our rather eccentric physics teacher decided he had done enough coursework and could do things that interested us'.

John Yudkin, the nutritionist, asserts that 'it doesn't matter if you are taught physics or chemistry or biology. All pupils need is a good teacher who is enthusiastic about his subject'. Such was the quality that he discerned in his own biology teacher at Hackney Downs School: 'he created enthusiasm'.

they will be incorporated as an integral part of the normal course. It consists of stimulus activities, such as visits to scientific establishments or talks and lecture demonstrations from leading scientists and engineers. And it also consists of student research projects, including individual or group research, competitions, great egg races, and school–industry schemes. It is on these that most of this book will focus.

Student research projects

'Student research projects' is a term that I have coined to cover a range of investigational activities in science. They may be short or long, they may be done individually or in groups, they may be hypothesis-testing or observational, they may be purely scientific or technological and constructional. The key factors are that they should be focused on a problem of genuine interest to students and that students should take personal responsibility for the progress and outcome of the project. 'Student research project' is not an ideal name, it does not have an easy feel about it. Others have used terms like scientific investigation, exploration, projects, independent research projects, even great egg races. I would want student research projects to include all such which satisfy the criteria of students' interest in the problem and their responsibility for action. For such reflects the way scientists and technologists work, and thus enables students to develop, utilise and appreciate the methods of science for themselves.

For many students, and for many teachers, such projects form the crown of their science work at school. They enable them to produce work of the very highest standard, and give a sense of personal achievement which is invaluable in developing their self-confidence in and motivation to do science. They enable them to bring together many of their personal attributes, such as creativity, perseverence and cooperativeness, along with their scientific knowledge and skills. They give them the experience of learning what science is about, by partaking in the same type of activity as professional scientists. I am not advocating a

particular form of scientific methodology here: as Medawar (1969) observed:

> Why are most scientists completely indifferent to – even contemptuous of – scientific methodology? . . . because what passes for scientific methodology is a misrepresentation of what scientists do or ought to do.

John Ziman (1972) was closer to the truth when he said:

> Real scientific research is very much like play. It is unguided, personal activity, perfectly serious for those taking part, drawing unsuspected imaginative forces from the inner being, and deeply satisfying.

No wonder that students involved in such activity enjoy it, and find it highly motivating!

The debate about the nature and purpose of practical work in school science has been going on for a long time (see Layton, 1973; Jenkins, 1979; Woolnough and Allsop, 1985; Lock, 1988; Hodson, 1990; 1992; Gee and Clackson, 1992; Tytler and Swatton, 1992). Too often, however, the practice of school practical work has been determined less by any educational rationale than by the constraints of a 50-minute period, the exigencies of keeping 30 lively students occupied and the commercial pressures of the apparatus manufacturers. Looking back through the hundred or so years that science has been taught in schools, we find a succession of national and HMI reports deploring the sterile nature of much practical work in schools and the advocation of investigational project work. Too often practical work has been dominated and distorted by an aim to elucidate or discover some piece of scientific theory. This has caused both the practical to be a grotesque distortion of real scientific activity (by having to be too tightly structured to discover 'the right answer') and also, as recent Assessment of Performance Unit (APU) research has shown (Black, 1990), to lead to unsuccessful comprehension of the underlying theory (which has become too complicated by the 'distracting clutter' of the experimentation).

And yet, alongside this rather unproductive 'cookery book' type of standard practical work,

Box 3.3   Brian Woolnough, 'Investigations – not a "tame" type of practical!', editorial in **Physics Education, 23** *(1988)*

I had a dream. I was in school, teaching students how to paint. I taught them about the great painters, I showed them some of their great paintings, I even explained about the social and economic implications of art in the world of work. I taught them the basic principles of art, of composition and balance in a good painting, of the ways of mixing the primary paints to produce the required colour, and what types of brushes and canvases were appropriate for different types of effect. I taught them the range of skills required in painting, practising their brush work, ensuring that they could cover a sheet of paper with an even wash and could also paint within prescribed lines, I even ensured that they could produce a nice picture by giving them a painting-by-numbers exercise. To ensure that they were making good progress I assessed their work with a series of test items, checking that they could hold a paint brush correctly, could select and set up a canvas on a stand, and observe seven differences between a Van Gogh and a Turner painting. Then a student came up to me and asked when he could paint his own picture. 'Not yet,' I said 'you are not ready. You have not developed the appropriate skills. You must learn more of the underlying theory before you can apply it. It's too risky, you might make mistakes. Being a real painter is not a tame activity.'

I had another dream. I was in school, teaching students how to climb. I showed them photos of mountains, of climbers in action and explained how climbing had developed from being a necessity for those living in the mountains, to a sport for the leisured classes, until it had become big business and an important part of the economy for many countries. I taught them the underlying principles relating to snow and rock conditions, to the way that the atmospheric pressure decreased with height, and to the strength of ropes and how they were to be used. I gave them problem-solving activities, in untangling climbing ropes, and simulation exercises in planning a climbing expedition to develop the interpersonal skills so necessary on a mountain. I ensured that they had acquired the appropriate skills, and could demonstrate appropriate progression, by testing them in the gym with press-ups and rope climbing exercises, and graded them

according to standardised criteria on the number of press-ups and height climbed in one minute. Then a student came up to me and asked me when she could go climbing on the hills behind the school. 'Not yet,' I said, 'you are not ready. You have not developed the appropriate skills or learnt the underlying principles that you must apply. It's too risky, you might fall and hurt yourself. Climbing real hills is not a tame activity.'

I had a third dream. I was in school, teaching students how to do physics. I . . . . . . Then some students came up to me and asked when could they do some real science. And I woke up, and saw that many students in many schools were doing just that, in investigational practical work. The fact that such activity is not 'tame' – that it cannot be synthesised or assessed in a tidy way by teaching only preparatory theory and reliable, so-called component, skills – had not prevented an increasing number of teachers giving their students the opportunity of developing scientific potential through it. Though too little is known about the way that scientists actually attack their problems, or the way that students succeed with their investigations, it is recognised that personal attributes such as perseverance, creativity and motivation coming through challenge and commitment are fundamental. Experience, both tacit and explicit, coupled with self-confidence in facing new challenges can be developed through students tackling their problems in investigational practical work. Developing skills and acquiring understanding clearly have a place in science teaching, but we must ensure that it does not stop there. Students can easily under-achieve in their school science lessons, not because the tasks are too difficult but because they are too easy. Where scientific activity has been reduced to a series of trivial de-contextualised tasks it is not surprising that students lack motivation to succeed. We must ensure that physics teaching is not reduced to the equivalent of painting by numbers or climbing ropes in a gym. The investigational practical work described in this edition [*Physics Education*, 23(6), 1988] illustrates the high quality of scientific work that students can develop and demonstrate when encouraged to face real challenges.

some science teachers have been encouraging individual student research projects, not because they have been imposed by the external syllabus, but because the teachers believe that they are more representative of real science and consequently produce higher-quality scientific work from students (see Box 3.3). The tradition for student research work in school science, if only a thin red line, has been a long and noble one (Woolnough, 1988).

Correspondingly, many industrialists and educationalists have also advocated the virtues of giving students more initiative, of harnessing their creative talents, by doing more research investigations. Science teachers who have tried them advocate them because they work: the students are motivated, they enjoy them and learn science through them. Industrialists advocate them because they help develop the autonomy, creativity, problem-solving, teamwork, communication and entrepreneurial skills that are so important in the world of work. Educationalists advocate them because they fit in with psychological theories of how children learn best. They had a place in the progressive movement for education, after Rousseau and Dewey. Tytler (1992) quotes Kilpatrick describing the progressive movement which sought to base the curriculum around the interests of the child as aiming 'to build strong-charactered, social minded, self-governing persons' possessing the traits of 'self-respect, self-direction, initiative, acting on thinking, self criticism, persistence'. Clearly student research projects can do much to encourage these aims, though I would also want to argue, much more than many of the advocates of progressive education, that they need to build not only on students' interests (which can be limited and limiting) but also on a structured base of knowledge and experiences which need to be provided and directed by sympathetic teachers.

Perhaps the greatest, most influential characteristic encompassed in student research projects is the element of personal challenge. The learning, satisfaction, self-confidence and commitment that come through satisfying a personal challenge cannot be overemphasised, and are often missing from much school science. The trick is to provide such challenges that are stimulating without being too daunting.

In an important book, Hodgkin (1985) elaborated on this theme, expressing it in language that is accessible. He argues for an oscillating model for learning, influenced, I suspect, by his mountaineering background. He stresses that personal commitment and challenge, springing from a supportive and stimulating environment, are essential for personal growth and learning. He divides activity into three categories: 'play', to build up interest and basic familiarity (in the foothills); 'practice', to build up basic competences and confidence (on the training slopes); and 'exploring', to stretch, challenge and extend the frontiers (among the mountain peaks). He argues that a learner needs to be continually moving from one type of activity to another, to enjoy a wide range of experience – first playing in the foothills to stimulate interest, then practising to build up competence in the required skills, before facing a fresh and extending challenge in exploring new regions, and returning with renewed self-confidence to discover a new area in which to play. The affective aspects of interest, enjoyment, commitment, perseverance, self-confidence and challenge are fundamental throughout. Students need to be given freedom to play, practise and explore. The main task of teachers is to create a supportive environment, to take students into new areas in which they might play fruitfully, to provide the training in knowledge and skills which can then be utilised in challenging opportunities through their own investigations (see Figure 3.1).

The parallels in the three main types of practical work (Woolnough and Allsop, 1985) are clear:

- Play = practical experiences to build a feel for the phenomena and an interest in the area.
- Practice = practical exercises to develop competence in specific skills and knowledge.
- Exploring = practical investigations (or student research projects) to acquire stimulation, confidence and the ability to work as problem-solving scientists.

Fig. 3.1   An oscillating model for the creative cycle (after Hodgkin, 1985)

Russell Tytler (1992), in his splendid account of students' independent research projects in Australia; suggests that there are two educational premises underlying such child-centred and therefore interest-based activity:

> a learning theory that maintains that children can only really learn when they are interested in and value that knowledge which is being offered, and a theory of moral agency that maintains that the functioning of a democracy requires free and intellectually autonomous, hence knowing and purposeful, adults.

Tytler's study reports on students' projects done for the State of Victoria's Science Talent Search (STS). Such science fairs are an established tradition in Australia, and have been running in Victoria for 35 years. He studied some of the 4,000 entrants in the 1986 STS, first administering a questionnaire to 365 prize-winners and subsequently interviewing, in depth, a sample of the students to obtain, in their own words, the story of their projects and the nature of their experiences doing them.

The students interviewed were aged between 10 and 15, and it was clear that for many of them science projects were an important part of their thinking from one year's STS through to the next. Their projects covered an enormous range of subjects: designing and building a model hydroponic garden in an incubator; studying amino acid sequences in chicken embryos; modifying an electronic kit metronome; designing and building an automatic release/propellor start mechanism for a model plane; a study of spiders and the ecology of a local lake; measuring the physics of motion; the efficiencies and power output of model cars; the corrosion effects of chlorine on different surfaces; the effect of various household chemicals on growth in a herb garden; designing and building a hot air balloon with remote control elements; investigating the accuracy of weather forecasts over a period of time.

Various factors arose from this study.

- The academic ability of the students varied enormously.
- Many of the ideas came from random events in the students' own experience, their hobbies, their family or their reading at home.
- The commitment to the pursuit and conclusion of a project, as evidenced in the displays of enthusiasm and the sometimes extraordinary amount of time and trouble involved, was impressive.
- The independence shown in the pursuit of background knowledge and in the arrangement of experimental procedures or design innovations was also very impressive.
- The type of motivation differed from one student to another: for one a drive to build something; for another the urge to find out 'why'; for a third to develop an established hobby.
- The support of the home environment was very important, providing not only ideas and resources but also encouragement, interest and expectation. It provided the context for students' interests to develop in an environment where autonomy and independence were the norm.
- The amount of help needed and received from the school or teacher was very small.

- Interest and motivation rather than intellect are the key ingredients in pursuing a piece of research or a model to a successful conclusion; the key to this interest can lie within the home background or the school environment in which the projects are presented and pursued.
- The level of work that students are capable of and the range of talents they can bring to bear on science are enormous.

Tytler challenges us to find ways of bringing such achievement through largely home-based projects by a few students with unusual curiosity and independent drive into the context of the school curriculum for all.

We hope to show in the coming pages that this can be done and that, if we cannot actually *produce* creativity, we can at least *encourage* it in schools. We hope to show also that student research projects can be developed to some extent for all students, sometimes in small ways in normal lessons, sometimes through extended project weeks when the timetable is suspended, sometimes harnessing the 'creative spring' from their home environment with their school-based activity. (See Box 3.4).

Box 3.4   What children feel about 'doing science'

I have stressed the importance of the affective domain, the way students feel about their science, in the formation of their attitudes, their understanding and their career aspirations. The following extracts illustrate the way many students react to doing their own science in student research projects or other stimulus activities.

It was fun, especially when we had 20 pieces of paper and 10 cm of sticky tape. Mine went a bit wrong, but it was worth it. (age 10)

The lesson I enjoyed most of all was building the structure. I liked it because it was a challenge and also because our team won with the best structure. (age 10)

We had this wood stove at home . . . and I saw this white stuff coming out of the wood and I said 'What's this?' and my Mums says 'That's creosote'. So I collected some in a test tube and a couple of hours later I had it over a burner and I saw this flame coming out of the end of the tube and I thought 'This is quite volatile! This stuff is tar . . . quite incredible!' (13-year-old boy*)

. . . and this was particularly interesting . . . we were in the mountains using anti-freeze anti-boil in the car radiator and I was thinking of a substitute you could possibly use and so I tested alcohol and salt and a whole range of things . . . and I came up with some quite magnificent results for alcohol . . . when it was frozen, and the freezing point was lower when the alcohol was added . . . you got this magnificent branching effect on the top . . . I had no idea this was going to happen . . . branching is my name for it because it looks like a tree. (13-year-old girl*)

. . . it turned out the main thing wasn't the experiment itself but when I looked at the digestive juices underneath the microscope when I had nothing much to do I found all these little beetles and microscopic . . . beetles . . . and little eels and stuff like that all swimming around and I found about . . . 20 different sort of things just bobbing around . . . and that turned out to be the major part of my project . . . I think that was the most interesting part actually . . . It's actually like a little world. I can understand why people spend all their working day looking down microscopes now. (14-year-old girl*)

Box 3.4   What children feel about 'doing science' (continued)

I've learnt a lot . . . about hot air, aerodynamics, about the size of balloons . . . and I've been doing research about solar power and learning a lot about that too. I think that the best thing is the challenge . . . like you can't control where it [his hot air balloon] goes . . . so . . . (15-year-old boy*)

You can choose your own topic, and you find out things for yourself . . . you don't read it out of books, and you learn it and that's it. Instead you learn about something, and you learn it first hand. Like in the lake . . . water temperatures and the various insects that live in the lake, . . . and the spiders. (15-year-old boy*)

I enjoyed the investigation as I enjoyed the freedom of it. I could work at my own pace and do what I wanted. (16-year-old girl)

. . . So now I was all ready to start my test run [of a firework rocket]. I needed to do three things all at once so mum came out to help. She lit the touch paper and timed the explosion while I attempted to pull the paper through at a constant speed. The result of this experiment are shown overleaf. Not too impressive eh? and certainly not what we had expected. The rocket ignited and shot off destroying the springs and ripping the screw from its holding, only to be stopped by colliding with the buffer at one end . . . (17-year-old boy)

My own education was inspired by a teacher who introduced me to a scheme called 'Awards for Young Investigators'. The scheme encouraged you to experiment and think, but kept it at an interesting level. The scheme helped me use my imagination, an element that can so easily be lacking in teaching but which scientists need. (a physics undergraduate)

I also attended a lunch time science club where we were given a lot of freedom to explore and experiment. I remember investigating fruits and metal electrodes necessary to make the best battery, and making weird and wonderful mechanical models from high-tech lego. I really enjoyed this because I didn't feel any pressure to be producing a correct answer or explanation at the end. I also found this with 'egg races' we had: those were competitions, for example to see which teams could build the highest structure or catapult a marble the furthest using a limited range of materials, within a time limit. Many of the class, who were not normally enthusiastic in science lessons, enjoyed these. (another physics undergraduate)

What did have a significant influence (on choice of career) was the degree to which I found each subject interesting and enjoyable. It is for this reason that activities such as the great egg race can, and in my case did, have a real impact on a child's choice of his or her career. My interest and enthusiasm was sparked off by the opportunity to complete a practical task, set by a physics teacher, not using a calculator or reams of paper, but bits of stuff I had around the house. I got feedback from the teacher for doing what I loved to do outside school hours – playing with bits of junk. The competitive aspect was no great incentive for my teenage mind. What did matter was that I could build this thing that would go as far as I could achieve. The sense of achievement when you see a cart with two old gramophone records as wheels, cut away by your fair hands, still running after three lengths of the sports hall on just one elastic band, was a feeling that I rarely got from my school work in other subjects. I am grateful that I was exposed to the exciting side of science and technology when I had the chance to decide where I wanted my life to lead. (an engineering manager after a degree in mechanical engineering)

(I am grateful to Russell Tytler for permission to use his quotes, those marked with an asterisk, from his Australian study)

Characteristics of student research projects

Later we will look at examples and case studies of student research projects, developed both as in-class and as extra-curricular activities. At this stage it is worth noting some of the characteristics which distinguish this type of activity, and which point to some of the reasons for its success. Student research projects will have a large measure of openness in their development. Tamir has suggested that this openness can occur at different stages of an investigation: in the problem to be solved; in the planning and operation of the investigation; and in the possible solutions to the problem. Based on this he produces a four-way classification of investigations, depending on whether each stage is open – that is, left to the student to decide – or closed (see Figure 3.2). The projects in the Australian STS were largely of type 1, with the student choosing their own problem and their own set of procedures, and with a variety of possible 'solutions'. Unfortunately, many so-called investigations in many school science lessons are of type 4!

	Problem	Procedures	Solution
Type 1	*	*	*
Type 2	*	*	
Type 3	*		
Type 4			

* Open stage of investigation.

Fig. 3.2 Tamir's four-way classification of investigations

A student research project generally will have the following characteristics:

- It is based on a question of interest to the student – though teachers often make suggestions.
- The prime motivation for the work will come from the student.
- The student will have a considerable amount of autonomy and independence – the project will be 'owned' by the student, even if the teacher

makes some suggestions and raises tactful questions.
- The project is open-ended, with no single 'right answer'.
- The problem is difficult enough to provide a challenge for the student.
- It takes place over an extended period of time, not less than a week, so that reflection, discussion and modification can take place out of class time.
- It entails working with one or more other students.
- It is likely to use out-of-school resources; from home or the school 'junk box'.
- It normally involves adults other than teachers, providing advice, resources, stimulus or judgement.
- It builds on the student's existing knowledge and also information gleaned by personal research.
- It has an outcome, a report or an artefact, which is available for a public audience.
- It provides a successful and satisfying outcome for the student.

Specific criteria are provided for specific projects, but those enumerated above stress the generic aspects which ensure student satisfaction.

But the fun and the achievement in doing student research projects, as real scientific research, do not come easily or to the unprepared mind. In Pasteur's famous phrase, 'chance favours the prepared mind'. They need to go alongside, and build upon, a growing understanding of scientific knowledge, both explicit and tacit. Perhaps the following example will illustrate the relationship between theory and practice in project work.

I was teaching physics to a group of 16-year-old students in school. We had been studying the basic laws of mechanics early in the sixth-form course, Newton's laws of motion, momentum, impulse and so on, and decided to spend some time doing research on firework rockets, as we were approaching 5 November, Guy Fawkes day. Working in groups of two or three, and taking all appropriate safety precautions, some chose to measure the energy of a rocket by measuring its maximum height of trajectory with a simple, home-made theodolite; some chose to measure the calorific value of the fuel by making a simple 'bomb calorimeter' out of a film cassette holder; some measured the variation of force with time by mounting the rocket on a home-made frame and firing the rocket against springs; others measured the temperature of the exhaust gases with a simple thermocouple which they had made and calibrated. The motivation for and commitment to these projects, the students' own problems, were considerable and enabled the students to personalise the principles of physics through ownership in a way they would not forget. They wanted to know and use constant acceleration equations, and the relationship between heat loss and temperature rise. They reached a new understanding of the meaning of force–time graphs, with impulse and momentum gained being represented by the area under the graph, drawn on a moving piece of paper by a pen fixed to the firework rocket as it was firing. They learnt much about thermocouples as they researched them to meet their own 'need to know'. And, alongside the consolidation and development of the theory, they were gaining real experience of experimental techniques and errors as they sought to devise methods which gave accurate results, and would withstand the critical appraisal of their peers.

Knowledge without action is sterile; doing science without knowledge is trivial. The message of this book is not that effective science teaching is either all curriculum- or teacher-directed or that it is all student-centred, allowing students to discover the processes and content of science for themselves. It is both. I suspect that in the past we have concentrated too much on one or the other, emphasising either a soft progressive child-centredness or a hard traditional content approach. My message is that effective science teaching needs both; students do need to have gained a solid foundation of scientific knowledge and skills, which is determined by the teacher, but they also need to develop their own confidence, competence and intellectual strength by having the opportunity to take responsibility for their own extended research projects in science.

CHAPTER 4

Student research projects

We have seen something of the genesis of student research projects and the characteristics that make them so effective. It is possible to fit such projects into the normal curriculum of science courses, either as part of the formal requirement of the course, or as optional part of the assessment, as in Midlands Examination Group Double Award Science for GCSE, or brought into any course on the teacher's initiative as being appropriate for students. However, because of the restrictions and constraints of the normal science curriculum, with the limitations of time and space as well as the pressure to include more science content, many teachers have found that they have been able to encourage students research projects best through extra-curricular activities such as science clubs, competitions and CREST awards. In this chapter we will consider examples and different ways of doing student research projects, both in-class and extra-curricular.

In-class student research projects

The Scientific Investigation requirement in the National Curriculum for science in England and Wales (DES, 1991) encourages many of these same research characteristics, especially for older students. At Key Stage 4 (ages 14–16) the Programme of Study requires students 'to communicate, to apply, to investigate and to use scientific and technological knowledge and ideas to make informed judgements' and 'to articulate their own ideas and work independently or contribute to group tasks. They should develop research skills through selecting and using reference materials and through gathering and organising information from a number of sources and perspectives'. Students are 'encouraged to develop investigative skills and understanding of science activities' which, among other things, 'promote invention and creativity', 'encourage planning and evaluation', '[encourage] the use of secondary sources', and 'may take place over a period of time'.

Some of the examples given, and some of the statements of attainment, suggest a rather more closed interpretation of scientific investigation, based on 'hypothesis testing' leading to the 'right answer'. However, others have seen open-ended investigations as more satisfactorily fulfilling the National Curriculum requirements, and have not been limited to such a pure science model. Earlier versions of the National Curriculum for science insisted that the higher levels (8–10) of Attainment Target 1 could only be achieved in the context 'of an extended investigation', and this still represents the spirit of the requirement. But that has now been modified to accommodate the practical difficulties of large numbers of students carrying out extended projects at the same time in already overworked laboratories with limited resources in school. For a fuller discussion of the role of Scientific Investigations in the National Curriculum see Gott and Duggan (1994). Richard Gott and his co-workers have been very instrumental for the introduction of such genuine scientific

Fig. 4.1   Developing and testing electric and electronic circuitry for alarm systems in a house

activity into school. Suffice to say here that the National Curriculum, with its inclusion of complete scientific investigations as an integral part of science, has done much to encourage student research projects in schools – even if its reductionist assessment structures and its administrative overload have done much to make their introduction more difficult.

The 5–11 age group

Science in primary schools traditionally centred around the nature table and studies of the environment. Through the 1960s and 1970s, however, various initiatives were introduced to strengthen the experience that children had of science before they entered secondary school. Nuffield Junior Science (Nuffield, 1967) and the Schools Council *Science 5–13* (Ennever and Harlen, 1972) produced some excellent, if rather wordy, guides for teachers. However, many primary schools teachers felt ill at ease teaching science as they were conscious of their own lack of scientific education. Much work was done by local science advisers, teacher trainers and the Association for Science Education (ASE) to build up teachers' confidence and competence. The sheer number of teachers in primary schools, all of whom would be expected to teach science, meant that such a task was immense.

Inevitably, different approaches and schemes of work were proposed to the teachers, with science portrayed either as a process or a body of knowledge. Many of these emphasised science as a way of looking at the world, as a process, as a way of systematically investigating and finding out about the natural and the created world. Some excellent student investigations resulted, with teachers and students finding that science held no terrors but was an extension of the investigational approach used to cover other topics. Other approaches, however, stressed the content side and saw science in primary schools as consisting of pupils learning scientific knowledge, a watered-down form of secondary science. Unsurprisingly, some teachers found more confidence in science as a closed body of knowledge, which they could learn and transmit, than in the more open-ended investigational approach. When the National Curriculum was introduced, with science as a core subject for all pupils from the age 5 to 16, science was spelt out in meticulous, inappropriately precise, language which stressed both the content of science (in ATs 2, 3 and 4) and the exploratory and investigational nature of doing science (in AT1).

There are, undoubtedly, practical problems in incorporating extended, open-ended projects for all pupils in school. But there are opportunities for introducing shorter, more constrained projects into normal classroom time, and these have been

Box 4.1   *The strongest conkers*

Alistair Ross explains how, early in the autumn term, his class of 9 and 10 year olds at Fox Primary School, London, tried to find out which conkers were likely to be the strongest.

The children had many ideas, Sam thought . . .

The Conker should be big or small but if it is in the middle it is not so strong. The Conker is also better if it is older.

Is this true? We collected over 200 conkers, including part of a precious hoard of year-old ones. We numbered each one, and recorded the day the new ones were collected from the park — we assumed that they had fallen from the trees the same day.

We weighed each conker, measured its strength in The Destructor below, and then the thickness of its skin with a micrometer screw gauge.

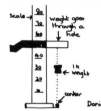

A launching platform slid up a 2m scale.

We dropped the 1k weight from increasing heights until the conker below cracked — we called that height (in cm) the strength of the conker.

Daniella

Sorting out our results

We made an enormous chart of our results, but could not see any patterns in them.

So we made a data-file on our microcomputer * and used that to answer our questions.

Pattern in the weights

We asked the computer for all the weights of the conkers starting with the lowest weight and then getting heavier

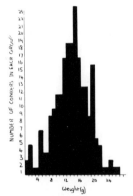

"Most of the conkers were in the middle and most were between 10g and 18g. Only a few were at the edges." Martha

─────────

* We used the micro-LEEP package on an RML 380Z machine.

The FACTFILE package, sent to schools with their MicroPrimer pack, could be used to do similar (though less extensive) work with cassette-tape drive machines.

The heavier the stronger

This is part of the computer print-out which gave us the data for the graph.

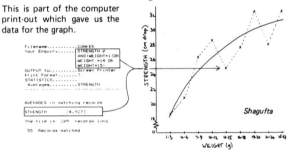

Shagufta

The younger the stronger — at first

This graph shows that as a conker gets older it gets weaker and then sometime in the year it starts getting stronger again

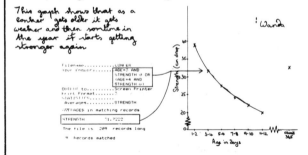

Wanda

Our conclusions

We drew many other graphs and charts, and finally agreed that:
- the newer the conker the stronger it is likely to be — unless it is kept until next year;
- heavy conkers are stronger than medium ones, which are stronger than light ones;
- large conkers are usually stronger than small ones;
- the thicker the skin the better — though as this could not be seen from the outside it did not help in the choice of a good conker!

encouraged and developed very well for younger pupils when the pressure for content coverage has been less severe. In the primary and junior schools (ages 5–11) projects such as *Science 5–13* (Ennever and Harlen, 1972), ASE Primary Science leaflets, and BAYS Young Investigators have been influential and many individual teachers have developed genuine scientific research work of a very high standard with their pupils, as has recently been shown at many ASE members' exhibitions (see Box 4.1).

The 11–14 age group

In the early secondary age range many students have been carrying out short, but genuinely scientific research projects around texts such as *Active Science* (Coles *et al.*, 1988–91). This popular course is content-based, but also has integrated exercises developing the scientific processes of communicating and interpreting, observing, planning investigations and, above all, investigating. Most of these investigations are relatively short and constrained and may only take a single or double period, though some will take longer and need time for homework (some examples are given in Box 4.2).

Box 4.2　Examples of Active Science investigations

Year 7　Investigating how the rate of breathing depends on age.
Making and comparing model roofs made from balsa wood, card and polystyrene squares.
What affects the voltage of batteries made from fruit and metal electrodes?
Year 8　Making and investigating a home compost heap.
Investigating parachute design.
Comparing the strengths of different fuels, diesel or solid.
Year 9　Investigating the size and shape of craters (using marbles and a sand tray).
Making and evaluating the effectiveness of a light-dependent resistor-based burglar alarm.
Investigating how the amount of yeast and sugar, and its temperature, affect fermentation.

Though students have little choice of the individual topic, which is set into the context of the scientific theme being studied, the problems are sufficiently open-ended to allow students to devise their own methods and thus acquire personal ownership of the project. They do give students a good insight into, by involvement in, the way scientists work and a feeling of ownership of the project, because they have done the planning.

The GCSE stage

It is at the GCSE stage, at ages 14–16 that the pressure on work for the 16+ school leaving examinations has been greatest, and teachers have found it most difficult to fit in investigational science projects into their normal pattern of work. Prior to 1988 students were examined either by GCE (the ablest 20% of students) or CSE (the rest!). GCE examinations had no practical component, so teachers were left to do whatever type of practical they thought appropriate. Many did very little practical at all, many did routine 'cookery book' experiments to verify a principle or measure a property. A few, however, involved their students in genuine scientific projects and produced work of a high standard. Those following the splendid Nuffield Secondary Science course (Misslebrook, 1971) were also expected to do practical investigations.

When GCSE was introduced in 1986, bringing together the GCE and CSE examinations, a common pattern of practical assessment was introduced for which the students received up to 20% of the marks. This assessment was based on objectives from the National Criteria of the Secondary Examinations Council, and related to 'the assessment of practical and experimental skills'. These were slightly different from the separate sciences, physics, chemistry and biology than for science, but all followed a similar pattern. These criteria analysed what was being done in scientific experiments and reduced this to 15 component skills (see Table 4.1)

It was a reductionist approach and emphasised the parts rather than the whole. Nowhere was the

Table 4.1 Objectives from the National Criteria (GCSE Science) which relate to the assessment of practical and experimental skills

Candidates are expected to demonstrate the skills and abilities to

- Observe, measure and record accurately and systematically
- Follow instructions accurately for the safe conduct of experiments
- Communicate scientific observations, ideas and arguments logically, concisely and in various forms
- Translate information from one form to another
- Extract from available information data relevant to a particular context
- Use experimental data, recognise patterns in such data, form hypotheses and deduce relationships
- Draw conclusions from and evaluate critically experimental observations and other data
- Recognise and explain variability and unreliability in experimental measurements
- Devise and carry out experimental or other tests to check the validity of data, conclusions and generalisations
- Devise and carry out experiments or other tests for particular purposes, selecting suitable apparatus and using it effectively and safely
- Apply scientific ideas and methods to solve qualitative and quantitative problems
- Make decisions based on the examination of evidence and arguments
- Recognise that the study and practice of science are subject to various limitations and uncertainties

Source: From Joint Council for 16+ National Criteria (1981)

idea of a scientific investigation stressed, though one did speak of the need to 'apply scientific ideas and methods to solve qualitative and quantitative problems'. The examination boards followed this reductionist approach, though tended to group the practical skills around more general headings such as observation, manipulation, design and planning, recording and reporting. Each of these broad skills had to be assessed separately, a number of times, against predetermined criteria statements. All this put a great emphasis on practical work, assessed on its component parts by the teachers.

The process of practical assessment, with its structured experiments, its administrative recording, and its cross-moderation caused much time to be devoted to a particular type of structured practical work which discouraged teachers and students from taking on genuine scientific investigations in their GCSE work.

There was at this time a considerable debate about the best way to assess and the best way to teach practical work. Some took a reductionist approach, stressing the component skills and the reliability of the assessment; others took a more holistic approach, stressing the need for students to do whole scientific investigations and the validity of the assessment (Bryce and Robertson, 1985; Woolnough, 1988; 1989).

In the event, the National Curriculum for science opted for a more holistic approach to practical work and stated that one domain in science should be Science Exploration or, in the later versions, Scientific Investigation. This should be given 25% of the marks at Key Stage 4, the GCSE level. Although this Scientific Investigation was split into three strands (asking questions, making predictions and hypotheses; observing, measuring and manipulating variables; and interpreting the results and evaluating scientific evidence), it was stated explicitly that the assessment of these strands must be done in the context of doing a whole investigation. Thus the National Curriculum encouraged student research projects through the assessment of this Attainment Target of Scientific Investigation. However, it should be added that the pressure on time caused by the large knowledge content to be covered in National Curriculum science means that many teachers find it difficult to do long student research projects, replacing them with short investigations instead. There are still problems with this Attainment Target, with an ambiguity between the investigations of the Programmes of Study and the hypothesis testing in the Statements of Attainment (Tytler and Swatton, 1992), but the overall message is one of encouraging teachers to take a more holistic approach to practical science through doing scientific investigations.

All of the GCSE boards will now allow extended projects to be used as one form of teacher assessment for this Sc1 Attainment Task; indeed, they encourage them as being in the spirit of the National Curriculum. Currently, however, the Schools Examinations and Assessment Council insists that some evidence of scientific investigational skill must be demonstrated in each of the three scientific knowledge Attainment Targets, in the biological, the chemical and the physics areas. So, if an extended project is done in, say, a biological area, in Sc1, evidence must also be demonstrated in an investigation from chemistry and physics, Sc3 and Sc4, too. So, although a long investigation will be encouraged, if it encompasses work in only one of the three sciences it will need to be supplemented by (short?) investigations in the other two. At the time of going to press (1994), the Sc1 Attainment Tasks of the National Curriculum in England and Wales is being rewritten to make it simpler and remove the more obvious problems contained in the 1991 version. Hopefully it will still encourage scientific investigation, but in a less prescribed way.

The A-level stage

Perhaps the most obvious place in the school curriculum for student research projects is at the post-16 stage when more mature, abler, more committed students have chosen to study science in more depth and devote more time to it. These students have a greater amount of scientific knowledge behind them. There have, indeed, been many examples of such projects being done at this stage, perhaps through fieldwork, in the laboratory or with an industrial link, though they still do not form part of the majority of sixth-form courses. Some subjects, such as A-level Nuffield biology and physics, UCLES Modular Sciences and the European Baccalaureate, do have a compulsory project or investigation as part of the course, and these have a considerable influence on the students who take them.

In my own subject, physics, about 100,000 students have already done investigations since the

Nuffield scheme was introduced in 1970, producing work of considerably higher standard than would be produced by standard A-level practicals (I write as a sometime moderator for the Nuffield A-level physics investigations). The best ones are quite exceptional in their quality. More recently, as a tutor in science education at a university department of education I have occasion to interview postgraduate physics students to see if they would make suitable science teachers and be acceptable for our postgraduate teacher training course. Usually I ask them to tell me something about their own experience of school physics and I can predict, in a way that would be dulled by repetition if each were not so personally exciting, that those who have done a project at school will enthuse about it as the most important and most memorable part of the whole of their school science course.

It has been said that doing projects in the sixth form after students have made their choice to specialise in science is like preaching to the converted, since their motivation to become scientists is already here. But this is not necessarily true. Many students doing sixth-form sciences have not made a commitment to a career in science or engineering: that will be determined one way or the other through the two years of the 16–18 science course. Many grow bored and decide against any further involvement with science or engineering at this stage. There is evidence that many who decide on such careers do so through involvement in a student research project as part of their A-level studies. But, of course, many students studying science at this stage have no intention of continuing it into a scientific career and any justification for such projects must be made equally for them. Fortunately this is not difficult, as the skills and self-confidence developed through such work are indeed appropriate to a general education. There has been much talk in the UK recently about the importance of core skills for all students studying in the 16–18 age range, skills such as problem-solving, personal skills (for instance, being able to work together in a team), communication skills (being able to present a

Fig. 4.2   Using multishot photography and ticker timers to analyse the motion of weight-lifters

public report on a project), and IT skills. All of these have been an integral part of such project work since it was introduced!

The Nuffield A-level biology project sets out to develop and assess five practical skills:

1 Statement of the aim of the project.
2 Investigation and use of background knowledge.
3 Planning and carrying out the procedure.
4 Inferences from investigatory practical work and relating these to background knowledge.
5 Evaluation of practical procedures and suggestions for further work.

The project must involve an independent practical investigation of a problem and candidates must submit individual reports based on their own investigations. The teacher assesses the candidates' work against the above criteria, and his/her assessment is externally moderated.

The Nuffield A-level physics course calls its student research project an investigation, which it hopes will develop 'the ability to explore and investigate, to frame and test hypotheses, to recognise and solve problems, and to overcome or outflank the snags presented by awkward reality'. It applies its assessment to five broad criteria, against which the teacher makes an assessment. Again the teacher's assessment is externally moderated with the written report and teacher comments. Students are encouraged to write up the report in the form of a diary.

The assessment criteria are related to the following:

1 Applying knowledge and understanding of physics of an appropriate standard at all stages of the investigation.
2 Showing sensible scientific behaviour.
3 Devising and carrying out simple, effective experiments.
4 Incorporating a variety of worthwhile ideas – even if they are not always successful.
5 Appreciating the meaning of observations and being able to evaluate them critically.

The UCLES Modular Sciences course incorporates in its Extended Study module 'two, three or four extended investigations one of which should, if possible, be related to work experience, industrial visits etc.'. The different number of the investigations permitted allows for investigations of different type, depth and length to be undertaken, according to local circumstances. This module provides the opportunity for students to carry out extended scientific investigations, to communicate by means of extended prose, to look at science in a working industrial or economic context and to gain (further) experience of the 'world of work'. The practical assignments, whether laboratory- or field-based, are assessed against five qualities:

1 Identifying, defining and planning the investigation.
2 Research, use and understanding of relevant information and concepts.
3 Carrying out of experimental work with due regard to safety.
4 Recording of results and presentation of the assignment.
5 Interpretation and conclusions.

Each of these qualities has three 'performance criteria descriptors' to guide the teacher in making an assessment on a six-point scale.

The range of investigations and projects in these schemes is enormous; some of the apparently simplest ones often turn out to be the most productive. Students are encouraged to decide on their own problem for investigation, though teachers may want to discuss exemplars from previous students' work, or ideas from the examination boards, and use their professional judgement and tact to discourage what is likely to be impractical within the time and resources available. Most of the projects will arise from the normal coursework in the science lessons though others may arise from students' own interests or from problems on a local company's agenda. The examination boards administering the respective examinations are able to supply further details and support for teachers doing investigations in these courses.

The value of such projects lies, again, in both the cognitive and the affective domains. They allow students to learn and understand more of the science in the course and, by using some of that knowledge in problem-solving, will personalise and deepen their understanding in a way that second-hand experience cannot. Furthermore, the students will develop self-confidence and important interpersonal and communication skills, abilities which are truly transferable. They develop in students the autonomous approach to work and problem-solving which is so much in demand in 'life after school'.

Alongside the academic A-level courses in England and Wales, are being developed courses of a more vocational bias. These are being coordinated under a General National Vocational Qualifications (GNVQ) scheme. The scheme intends to coordinate all vocational, artistic, professional and academic under the same framework so that, for instance, a GNVQ would be equivalent to an A level. The assessment is based on competency statements and continuous, coursework assessment and while, at the time of writing (1993) the science course is still in the pilot stage, it is hoped that student research projects, probably linked to industrial placement, will be an integral part of the course.

Fieldwork

Probably the best-established type of student research project in the school science curriculum is that of fieldwork. Used predominantly in biology, though increasingly also in geology and astronomy, fieldwork can deliver all the main aims of practical work. It can develop practical skills through directed exercises to observe, analyse, measure and record the environment. It can give a feel for the phenomena being studied as plants, animals, rocks, fossils and environments are studied. It can develop the ability to be a problem-solving scientist as aspects of the environment are studied and variables investigated. Above all, it can develop the all-important affective aspects, the excitement, the wonder, the thrill of discovery, which are such motivating factors for young people. Whether it is a motivation which directs a student towards a career in science or which gives a future citizen an enhanced wonder and enjoyment of the world in which he/she lives, it is a motivation which little else in school science can deliver. The sense of wonder that a group of hardened graduate scientists showed recently when on a geological field trip they broke open a rock and found a fossilised shell – 'no-one has ever seen this before, ever!' – was wonderful to behold. Most students will find that a knowledge and understanding of their world will heighten their enjoyment of it and their awareness of their significance in it, as they walk through the countryside, study a cliff face on the sea shore or gaze at the stars at night.

Early on in my science teaching career I took a group of 13-year-olds to do some residential fieldwork in Wales. Each student tackled a specific investigation, focused on the wildlife, the flora and fauna of the environment, the shape of the land, the flow of the river or the path of the river bed. My most abiding memory was of the quality of the students' work, which was considerably higher than the work they had provided for me in my normal science lessons back in school. And the difference was the students' motivation and commitment to their task. Because they had been given the opportunity to select a project, and the responsibility to tackle it their way, they had taken ownership and seen problems as a challenge rather than a barrier, and the presentation of their results was something in which they took a pride rather

than a teacher-imposed chore. Of course, the time and the context made a difference – there is no way one could replicate such activity in one-hour science lessons back in school – but the principles of ownership and personal challenge, of not underestimating the quality of scientific work that young people can produce, are certainly transferable, if unhindered by constraints imposed by the syllabus, by assessment strategies or by national curriculum.

The principles of successful fieldwork are fundamentally the same as for other types of practical work.

- Students need to be prepared, they need to know what they are looking for, what are the norms (so that abnormalities can be noticed), and how to observe, measure and record appropriately.
- The problem needs to be focused, but open.
- Students need the opportunity to develop their own approach, if possible to decide on their own investigation, so that personal ownership can be achieved.
- The task needs to be manageable so that students achieve success in it, but challenging enough so that they feel stimulated.
- The students need to be clear about what they need to do and how results will be recorded in rough note books, prepared tables, sketches or photos, in diaries or logs, as well as in final reports or displays.
- The activity needs to be well prepared and tested, with clear instructions, adequate resources, accessible reference books and guides, and, of course, explicit safety guidance.

This is not the place for an exhaustive discussion of the possibilities of fieldwork in school science. Suffice to say that the scope of such work can be as varied as the imagination and resources allow. At one extreme, residential courses to exotic places enable different locations to provide new stimuli. At the other extreme, studies of students' own gardens, the school grounds, even some unpromising wasteland nearby, can provide the opportunity for student research projects which can make science teaching more effective. Though even local fieldwork can rarely be fitted into a single period it does provide the opportunity for students to carry out long-term investigations in their own time, out of school hours.

Extra-curricular student research projects

Young Investigators

To help stimulate the teaching of science in junior schools the British Association for the Advancement of Science introduced its Young Investigators award scheme in 1982. This was intended 'to help young people to develop their sense of curiosity into an ability to investigate in a systematic way and thereby to learn and enjoy the basic approach of science and technology' (British Association for the Advancement of Science, 1982). The awards are designed as an 'out-of-school' activity for young people between the ages of 8 and 12. There are three levels of awards: bronze, silver and gold. The scheme is non-competitive, the pupils measuring themselves against certain general, explicit criteria. The criteria of success – and it is the firm aim that all pupils should achieve success in their project before being assessed – are made on the professional judgement of the teacher and another independent assessor. The scheme aims to build up enthusiasm for and sense of personal achievement in doing science, and has already achieved enormous success. In 1992 over 20,000 pupils gained YI awards, 80 per cent at the bronze level, 18.5 per cent at the silver and 1.5 per cent at the gold.

To gain a bronze award pupils are required to complete ten tasks, some prescriptive, some open-ended and some problem-solving. It is expected that the tasks should be spread over three terms, covering topics such as materials, ourselves, colour, living things, flight, the sky, air and water. They are intended to help young people learn:

● skills and care in manipulation of tools and materials;

● to be patient and keep records;
● to tabulate data and use tables;
● to apply different methods of measurement and investigation of living and non-living things;
● that the ways in which materials are used are influenced by the materials' properties;
● to predict the effect(s) of certain changes through observation of similar changes;
● to observe objectively, and take safety precautions;
● to choose suitable means for the recording of observations and the expression of results;
● and that more than one variable may be involved in a particular change.

After achieving bronze awards pupils may go on to take the silver award, a project taking about 20 hours' work (perhaps one hour after school for 20 weeks), and even, perhaps, a gold award, for which 30 hours' work is expected. A network of local coordinators has been established throughout Britain to give support to teachers and pupils, to help with adjudication and to make the presentations. Supporting material in the way of notes for leaders, suggestions for projects, exemplar case studies and assessment criteria are provided by BAAS. The assessment scheme outlines the general criteria required and is carried out seriously, but in a 'low key and relaxed' manner by supportive and enthusiastic adults. The children delight in talking about their work with such sympathetic adults, showing them their investigation, their log book and presentation display, and gain an enormous degree of satisfaction from the completion of their projects. Such projects, because they address the pupils' own problems and expect them to use their own initiative to resolve them, have a highly motivating effect. The satisfactory completion of the project, with the associated public award, does much to build up pupils' self-confidence in science.

CREST awards

Following on from the success of the Young Investigators award scheme, the British Association, in association with the Standing Conference on Schools Science and Technology (SCSST) and

Box 4.2

CASE STUDIES

SILVER AWARD

JULIAN'S PROJECT

Julian had always been interested in birds and in doing this project he saw an opportunity to find out for himself rather than always getting information from books.

However a certain amount of research was necessary, finding as much information as he could about the birds which would be likely to visit his garden. He wrote to various organisations such as the RSPB, the Wildfowl Trust and the Lincolnshire and South Humberside Trust for Nature Conservation. In this way Julian collected much valuable information about when and how to feed the birds.

With the help of his father he designed and built his own bird table and found a suitable site for it in the garden. In order to ensure a varied supply of food, Julian designed and built feeders to hang on the table. These designs were modified in the light of experience. These feeders were also used to devise simple tests to see how quickly the birds, blue tits in particular, could solve the problem and claim their peanut reward.

This work was carried out alongside other investigations into the feeding habits of birds. Initially Julian needed to discover what the different types of birds ate, when they fed and how the fed. This survey was carried out over several weeks and the observations were recorded on graphs and pie charts. With this information Julian was able to target his investigation at certain groups of birds.

A second bird table was then erected in the garden, designed specifically to attract the smaller, more timid birds rather than the quarrelsome starlings.

Julian now began an investigation into whether birds prefer one colour of food to another. Having discovered the popularity of fat and bacon rind, Julian used food colouring to dye lengths of each food. As predicted the uncoloured fat proved most popular still, but Julian discovered that different birds did appear to prefer different colours, possibly due to their relationship to the colour of their natural food, for example, red berries.

This investigation was continued using coloured pastry shapes and finally with coloured feeders made from milk cartons.

Julian realised that only tentative conclusions could be drawn from his relatively short survey. The winter was very mild making the feeding season a short one and Julian was forced to give up his investigation in the early spring when the birds began to nest.

British Association for the Advancement of Science

supported by the Department of Trade and Industry and various leading industrial companies, introduced their CREST (CREativity in Science and Technology) award scheme for secondary school and tertiary college students in 1986. The scheme aims to 'stimulate, encourage and excite young people about science, technology and engineering through task orientated project work', and this it surely does, encouraging students to produce project work of a very high level. One of the secondary aims of CREST is to develop constructive links between school and industry, and this adds a further motivating dimension as local engineers work with the students and their teachers to give a sense of realism to school science and technology (West and Chandaman, 1993).

There are three awards – bronze, silver and gold – each of which encourages self-assessment, self-motivation, communication, problem-solving and creativity. On average the amount of work required for the bronze award is 10 hours, for the silver 40 hours, and for the gold 100 hours, some of which will take place in the student's own time. The gold award project must be industry-linked. Though the awards are free-standing, producing their own profile sheet, badge and certificate, the projects can also be incorporated into the National Curriculum schemes for science and technology, and produce a very welcome portfolio for discussion at interview for industry or higher education. Some schools are incorporating CREST science projects into their GCSE Double Science course, which counts for the 25% internal teacher assessment. It is now possible to do such assignments as the National Curriculum moves into GCSE; it would certainly help to make sense of the Scientific Investigation of AT1.

Once again, it is the students' autonomy and personal responsibility for the project, whether working individually or in a group, which gives them such a motivating sense of personal achievement. Talking with the students as they are working on their project or when they are giving their final presentation through posters, photographs, diagrams and project report, immediately convinces the listener of the enormous sense of achievement and personal pride which the students have achieved through being real scientists and engineers on their projects. It is their projects, their problems, their ideas, their research, their investigations leading to their conclusions and their report which make such work so influential and memorable. Many years after leaving school, it is such extended research projects which are remembered with enthusiasm when all else has been forgotten.

The range of topics covered for CREST awards is limited only by students' curiosity and imagination. Perhaps the following list of recent project titles will give something of the flavour:

CREST bronze awards:

- The prevention of trailer theft
- Cleaning the skin
- The design and making of a garden tool for use by a handicapped person
- The investigation of a local habitat
- Factors affecting the viability of stored wheat seed exposure to heat and moisture
- Sulphur dioxide as a pollutant
- Examining the performance of simple electro-chemical cells
- Designing and making a light meter
- The variation of flow rates through taps.

CREST silver awards:

- The cellular and bacterial content of milk
- The effect of man's feet on his environment
- Alternative energy supplies
- A study of burnt heathland and its recolonisation
- The variation of soil temperature with depth
- The effect of water softening on detergents
- The effect on viscosity of stirring a non-drip paint
- The design and construction of an electronic weather station
- Enzyme activity in yeast
- The physics of athletics and athletes.

CREST gold awards:

- Automated bird photography

- Development of a pressure-sensing crowd control barrier
- Abrasion damage to leaf surfaces
- Designing and building a low-cost autopilot for a yacht
- The prevention of rancidity by packaging film
- Creating an electronically controlled environment for growing lettuce
- An investigation of ways of reducing damage to warehouse stacking facilities
- The testing of chromograph slide valves under working conditions
- Investigating the effect of temperature on the degradation of vitamin C
- The effect of salinity changes on *Oscillateria aghardii*.

The CREST award scheme is becoming increasingly popular in the UK. By late 1993, 2,000 schools were affiliated to it, with 50,000 students having taken part and achieved an award at one of the three levels. Apart from the three level award scheme, CREST sponsors various annual programmes to stimulate challenges in different areas of the curriculum. The Environment Award programme, sponsored jointly with the *Times Educational Supplement*, encourages schools, in cooperation with local business, to undertake science and technology project work about environmental issues. In 1994 a Life Science programme is being launched, with Zeneca, to encourage projects in this area. Engineers from Ford are sponsoring another CREST programme focused on the automotive industry, and this will stimulate projects more in the area of physics and mechanical engineering.

The CREST schemes have recently received formal recognition and support at both school and government level. The Secondary Heads Association (SHA) has actively promoted them to all its schools and the Minister of Public Service and Science, William Waldegrave has encouraged industry to support it. In the UK Government's first white paper on science for 20 years, *Realising our Potential: A Strategy for Science, Engineering and Technology* (1993), the CREST scheme is specifically promoted with a view to increasing the public understanding of the vital importance of science, technology and engineering to the country's economic future. One of the particular initiatives that the Government proposes in it is

> Government support, with other partners as appropriate, for the successful Creativity in Science and Technology (CREST) Award scheme for task-orientated project work in schools, in order to build on its relevance to curriculum requirements and to increase both the profile of the scheme and the extent of its coverage of the schools in the United Kingdom.

Teachers can easily become preoccupied with and limited by the formal demands of the National Curriculum, and feel that they cannot find time for student projects. Such support from the highest quarters must encourage them to try to find time. Furthermore, the profile of problem-solving skills used in the CREST award scheme matches the particular requirements for the GCSE and the National Curriculum in both science and technology. Projects being done for a CREST award, far from being an addition to the requirements for GCSE and the National Curriculum of science or technology, can in practice be a means of fulfilling them. Credit transfer schemes are also being developed for GNVQ programmes.

One of the most interesting aspects of the CREST award scheme is its attractiveness to girls as well as boys. At the bronze level 55% of the awards are achieved by girls. It seems that the cooperative nature of the scheme, the focus on real life problems, and the encouragement to work at one's own pace and explore various approaches, all prove attractive to girls – who find such factors lacking in much main line science teaching. CREST's new GETSET (Girls Entering Tomorrow's Science, Engineering and Technology) scheme attempts to build on their positive response by encouraging them to do the higher awards and perhaps go on to find career commitment. In this scheme, girls who have completed the bronze award take part in a two day national event (the 1994 one was in Imperial College

Fig. 4.3 A team of girls planning how they will raise 25 litres of water from the floor to table level, in a GETSET competition

London) with further fun projects and competitions.

It is clear that students taking part in the projects for the CREST scheme, cooperatively both with other students and with scientists and engineers in industry, find the scheme highly stimulating and satisfying, and acquire good insights into the way real scientists and technologists work. Whether or not it motivates the students to enter careers in science and technology (as many of them do) the

Fig. 4.4 A CREST award team who designed and made a cat flap which was computer controlled, so that it only opened for the cat with a specific fur pattern

Box 4.3

CREST SILVER AWARD
PROFILE OF PROBLEM-SOLVING SKILLS

You and your teacher will meet on at least three occasions to talk about your project
At these meetings your progress will be assessed using the statements below. Each
goal or criterion must be met on at least one occasion during the project. Of course,
this does not apply if a particular criterion is not relevant to your investigation.

Name _____

Project Title _____

CREST Project Number _____ Individual ☐ Team ☐ Tick one

You can

Number of times assessed

	1	2	3	Optional		
Produce a workable idea/ideas in response to the problem identified.						
Find out information to help you with your investigation.						
Transfer science/technology ideas from familiar to new situations.						
Show originality in how you understand and interpret the problem set and in the ideas you have for solving it.						
Select/reject possible ways of doing the investigation and explain your reasons.						
Design a fair test and predict what results you expect.						
Attempt to control several interacting variables.						
Use your results to find out if they fit the idea being tested.						
Explain what your observations or results tell you in the light of what you are trying to find out.						
Attempt to cross check your results and explain the results that do not agree with the predictions you made.						
Alter and improve your investigation in the light of what you have found out.						
Work carefully and accurately.						
Record different ways of doing your investigation explaining the strengths and weaknesses of each approach.						
Record and explain why there may be more than one explanation for what you found out.						
Understand in what ways your experiment or prototype might not work in a different situation i.e. prototype uncertainty/accuracy limitation of experiments.						
Explain how far you got in your CREST Award project, what problems you found and how you tried to overcome them.						

Signature of Student _____ Date _____

Signature of Leader _____ Date _____

FINAL EVALUATION

The two sets of criteria are the goals
which each student is meant to achieve
during a project. It is the role of the
Evaluator to make sure those goals are
relevant to the project have actually
been met.

The first set of criteria have been
used by the group leader to complete
the Profile of Problem-Solving Skills. The
profile will be used when the Final
Evaluation is made.

The Evaluator will decide whether the
second set of criteria have been
achieved at end of the project.

You should expect to

Explain clearly your part in planning and
carrying out the project using your own
words.

Explain how problems were overcome and
alternative solutions reached.

Produce a clear and concise record of the
project using technical language and style.

Demonstrate the fitness for purpose of the
product of your work. Your log will help you do
this.

Suggest where appropriate, potential
industrial/social/commercial applications
of the project work.

Signature of
Evaluator _____

Date _____

Box 4.4   Two examples of CREST work

Abrasion damage to leaf surfaces (Abstract)

Our work involved estimating the average percentage wind damage to Sycamore (*Acer pseudoplatanus*) leaves on three sites in the north of Ireland. One was an exposed site, one a sheltered inland site and other a roadside site. At each site we used ten trees and took twenty leaves from each canopy. We then compared this data with existing data from a variety of other sites in the same geographical region.

The percentage wind damage was estimated using a protocol previously established. The results revealed a very high level of damage on the exposed coastal site which could probably be explained by the large turbulence effects produced by the topographical features. The inland sites showed very little wind damage and this may be due to their low elevation and being on the leeward side of the east facing river bank.

Our values for wind damage to trees on the roadside site was unexpectedly low but the above reasons apply at this site also. Investigation of percentage damage against total leaf area shows a negative correlation for each site, this could be explained in one of two ways:

(i)   leaves which are heavily damaged may drop off prematurely or
(ii) as leaves become increasingly damaged this may have an effect on their growth.

The damage to the leaves could:

(i)   expose them to attack by disease or environmental pollution
(ii) cause the tree's growth to suffer.

This wind damage factor could have an adverse effect on the growth and production of a wide range of wild plants and crops.

The most damaging type of wind is non-laminar (turbulent) flow. It would therefore be advantageous to try to prevent turbulent flow if possible. Turbulence is produced by solid barriers e.g. buildings, walls and solid fences.

The baby's 'Clearway' project: from CREST to commerce

This student activity, founded in physics/science principles, translated into engineering design and manifested itself in a tangible product for manufacture.

The student developed a novel medical instrument (now called 'Clearway') based on his GCSE Physics project work.

The initial idea for the project arose through family discussions – both the student's parents are involved in the medical profession.

'The Royal College of Obstetrics and Gynaecology had just published a report on the hazards of using equipment to clear mucus from a baby's air passages immediately after birth. All the low-cost equipment on the market for mucus extraction had the possibility of bringing the midwife into contact with the HIV virus or with Hepatitis B.

'To avoid these dangers, midwives have reverted to holding babies upside down to clear their airways. If breathing does not start spontaneously, the baby is transferred for resuscitation to another room – further trauma for the mother. I decided to try to design a new piece of equipment which would eradicate these problems.

'My work resulted in a device that sucks the mucus away from the baby's airways by means of a small pump which is held and operated in one hand. The baby is held secure by the midwife with the other "free" hand.

'Made from biodegradable plastic, the device is easily disposable, is non-toxic and is therefore absolutely safe to use as there is no risk of contamination or cross-infection, and it can be produced for under £1.50. Furthermore, once used the pump can be broken off the device and a snap-on lid placed on the mucus container, which can then be labelled and sent off to the laboratories for analysis, if required.'

The student received a CREST Gold Award in 1990 for his work on 'Clearway'. He also won a Young Engineer for Britain category prize.

Around that time he became aware of the potential of the device and there appeared to be a real possibility that it would go into manufacture.

'Then the real work began. There is no easy route to turning a design idea into a commercial reality. There is a vast amount of research into every aspect of the product to be carried out and endless meetings which must always be matched by a determination to succeed.'

The student was able to use his Gold Award as a gateway, via the British Association, to the EC Young Scientists' Contest, where he won a top award and a COMETT Prize for Innovation. He also received £5,000 from the Comino Foundation's Fund to encourage commercial development of projects. On the basis of this he was able to proceed to setting up his own company to organise the manufacture and distribution of the product.

scheme certainly gives all students a greater understanding of science and increases their confidence and ability to live and work in their increasingly technological world.

'Great egg races'

Perhaps one of the most influential and certainly the most creative, initiative to have been introduced over the last decade or so has been the 'great egg race' competition. Whether it be management training courses for business executives or stimulus activity for primary school pupils, whether they last for days or a single period, whether they are tackled by teams or by individuals, the ubiquitous and ever-popular 'great egg race' has found a unique place in the educational system of England and Wales.

Though competitions testing creativity, perseverance, teamwork and other personal skills have been around for a long time, the 'great egg race' as such was probably started by the BBC in the late 1970s. This competition did indeed involve an egg: candidates – individuals or teams – were issued with a standard egg and a standard elastic band and told to 'construct a machine to transport a fresh egg the greatest possible distance using only the energy which can be stored in a small rubber band'. Certain rules were specified (British Association for the Advancement of Science, 1983) to ensure fairness:

1 The vehicle can be constructed from any material. It must self-start, it may not be given any

initial impulse or have any external guidance at the start or during the run.

2 Only the standard rubber band may be used to propel the vehicle. It may not be treated with chemicals, heated or cooled in any way. The elastic band may be cut but must be available for inspection as a single piece after a run.

3 The rubber band, as supplied by the competition officials, must be attached to the vehicle at the time of the run.

4 The egg as supplied must not be altered or added to in any way and must complete the run unbroken. It will be the judge's discretion to verify the weight of a contestant's egg at any time during the competition.

5 Distance travelled will be measured from the foremost point on the vehicle at the start tot the same point on the machine when it comes to rest.

6 Vehicles which are not ready to run five minutes after being called to the course will, at the discretion of the judges, be disqualified.

7 Vehicles which are unsafe or hazardous to spectators will be disallowed.

8 Anyone, an individual or team, can enter an egg-mobile.

Such rules are important as the degree of competition and personal commitment can be frustrated by a sense of injustice if a candidate's work is perceived to have been disadvantaged by an unfair technicality. Similarly, the criteria for judging, for selecting a winner, must be clearly specified in advance. The original egg race had a single

Fig. 4.5 The proud winners of a 'great egg race' to build a tower which supports three marbles more than one metre above the bench

criterion, the longest distance travelled. (The winner in the BBC competition travelled 183.3 metres, though the egg-mobiles of school pupils usually travel considerably less distance!). Many competitions also allow marks to be given for ingenuity, 'niceness' of design, even aesthetic value, and such are left to the judges' professional judgement.

Subsequently competitions with a range of tasks have been set, though still called 'great egg races'. Competitions to build the strongest bridge to span a given gap, with paper, or balsam wood, or spaghetti; to build the highest tower from newspapers to support three marbles or a flag; to build a satellite launcher which propels a table tennis ball into a given target; to build a boat (or a glider) which performs to given requirements; to make a vehicle which follows a line; to make a machine that sorts objects of different size, weight or constitution; to make an alarm clock which rings after 2 minutes; to build an automatic device which regularly waters plants (or feeds the pet) while the owner is away; to build a machine which, using given kitchen utensils and driven only by water power, will play a tune.

The range of such competitions is limited only by the creativity of the originator and the range of materials offered to the competitors. The British Association has produced two books of suggestions, *Ideas for Egg Races* (BAAS, 1983) and *More Ideas for Egg Races* (BAAS, 1985). The solutions offered by the competitors also demonstrate the abundance of ideas available when students are offered encouragement to be creative and when they take ownership of the problem and given from their own commitment. As Peter Medawar once said: 'creativity cannot be taught but it can certainly be encouraged'.

The organisers will decide in advance other criteria, such as time allowed and the range of material that may be used by the competitors, and again these must be specified and communicated to all competitors before the 'great egg race' starts.

Some teachers hold competitions in school time, either in normal class time or as a special event. Some encourage students to work on their design in their own time, and make it partly at home. This, of course, lays the project open to accusations of parental assistance which, while good in itself, may introduce unfairness in the competition. My own experience is that students working at home on their machines will be so possessive of their ideas and their creation that they will not allow their parents to interfere. Other teachers have avoided the problem, specifically encouraging pupil–parent cooperation by incorporating the competition as part of a PTA activity.

Those who have experienced such competitions will have been impressed by the enthusiasm, the creativity and the expertise demonstrated by students. They provide an opportunity for students to demonstrate their own ability and also to develop scientific problem-solving skills. Indeed, the way students tackle such problems is very close to the way practising scientists tackle their problems. John Ziman said in his BBC talk on *Puzzles, Problems and Enigmas* (Ziman, 1972): 'Real scientific research is very like play. It is unguided, personal activity, perfectly serious for those taking part, drawing unimaginative forces from the inner being, and deeply satisfying.' Many would say that that perfectly describes their experience in 'great egg races'.

The sense of occasion and excitement which can be produced by stage-managing the event with certificates, prizes, real scientists on the judging panel and generally showing that such a fun experience is taken seriously, all go to ensure that the competition is a satisfying and stimulating event for all concerned. Most students get a real sense of personal achievement from their involvement and satisfaction in having made a device which works even if it does not win. Many students would claim that they got their inspiration to become a scientist or an engineer as a career from involvement in such events. It is the affective factors, of motivation, personal satisfaction, self-confidence, challenge and commitment which are all-important in 'switching on' a student to a particular career.

Most 'egg-races' focus on problems in physics or technology. But this need not necessarily be a limiting factor. In fact, the Royal Society of Chemistry has brought together a splendid selection of ideas for chemical egg races, and other problem-solving activities in chemistry in a book

called *In Search of Solutions* (Davies, n.d.). They have similar aims and organisational structures, and have even considered the original egg race energy supply, the elastic band, and introduced the idea of a 'chemical elastic band' – 'anything you can do with a rubber band you can do with one teaspoon of bicarbonate and three teaspoons of citric acid' (*ibid.*). But they go well beyond that, building on genuine chemical concepts. They spell out their aims and justification for such activities in school science teaching clearly as the ability to

- enhance the understanding of the concept being studied. In this respect it acts very much as a reinforcing tool since it requires students to apply and, therefore, understand their acquired knowledge
- gauge students' understanding of their work . . .
- develop students' social and communication skills since it requires them to work in project teams
- build CONFIDENCE and aid MOTIVATION.
  (*ibid.*; capitals in original)

All of these objectives fit nicely into the structure and aims of the National Curriculum for England and Wales, including the practical Scientific Investigation of Sc1, which asks for collaborative and investigative work, adopting a holistic approach rather than individual skills. But they are not, of course, limited by that. The Royal Society of Chemistry expresses the hope of attracting young people into chemistry by showing them that it can be fun.

Each of their 'great egg' races starts with a task for the students, often set in a 'real-life' context, and is accompanied by guidelines relating to age groups, time duration, equipment required etc. for the teachers. Typical tasks are

- devise tests to compare various indigestion cures
- during the night a flytipper has dumped a load of waste in the local duck pond which has made it too acidic to support life. How can the pond be returned to a habitable condition?
- design and build an electrical battery which will give the greatest voltage, using materials only likely to be found in a kitchen

- design and make a boat which travels the furthest distance powered by one teaspoon of bicarbonate of soda and three teaspoons of citric acid
- your aircraft has just crashed on the desert and the last cupful of water has been spilled on the dry sand. You immediately scoop up the west sand and put it into a plastic bag. How can you get the vital water back? You have just 90 minutes before you die from dehydration . . .!
- make a 'chemical alarm clock' which goes off two minutes after setting, using a metal and a dilute acid
- find out how many different pigments are contained in a given collection of plant material
- determine how much available chlorine is present in four different bleaches, and which bleach is the best buy.

Safety considerations are particularly important in open-ended projects like these, especially those involving chemicals. Pupils should bring their ideas to the teacher for a safety check before starting on their experiment, but care should be taken to ensure that excessive concern for safety does not lead to teacher prescription, and limiting the pupils' 'ownership' of their own investigation.

Projects supported by industry and professional institutes

Over the last decade or so many companies and the professional institutes have supported and encouraged student research projects in schools, believing as they do that these both raise the quality of science teaching in schools and provide a highly motivating effect on all those involved. The number of companies supporting projects, formally and informally, locally and nationally, is vast. Many of these are now supported by the Neighbourhood Engineers scheme, described in Chapter 5. Traditionally, the education departments of big companies such as ICI, British Aerospace, UKAEA Harwell, and Unilever have supported such projects, but individual contacts in local industries have been no less effective.

The SCSST whose aim is 'to excite young people about science and technology, industry and engineering', also does much to encourage school–industry links. It supports a nationwide network of science and technology regional organisations (SATROs), which provide local support and supply individual links between schools and local industries.

The Sainsbury family, through the Gatsby Charitable Foundation, has been particularly active in this area, setting up a nation-wide Engineering Education Scheme. Its objectives are

- to ensure British industry has a continuing supply of high calibre, well motivated and qualified engineers
- to create a partnership between industry and education which fulfils companies' needs and individuals' potential
- to provide a succession of educational opportunities from school onwards to develop capability, leadership qualities and managerial skills.
  (Engineering Education Continuum, 1992)

This unashamedly aims to 'increase the number of high ability young people entering the creative levels of engineering' (ibid.), and targets groups of able lower sixth-formers in schools which it links with a local company. Together they select a problem which will be tackled in extra-curricular time by the lower sixth-formers and which will last from October until March. The net result will be a solution to the original problem, or at least a greater understanding of the problem, and a project report. It will also have given important insights into engineering to the students and developed their problem-solving and communication skills, their self-confidence and enthusiasm, as they get to grips with the realities of real problem-solving. The quality of much of the work done in this scheme has been very high indeed, and the tasks tackled very sophisticated; problems which are genuine ones for the company seeking to improve its current practice or tackle new contracts. They often gain gold awards in the CREST award scheme which are invaluable for student interviews for higher education. Many of the projects have also led to genuine engineering break-throughs, and commercially profitable outcomes (Clark, 1993; Hutchinson, 1993).

Currently, in 1993, there are nearly 600 students from over 100 colleges and schools participating, with about 100 partner companies on the scheme. Projects include redesigning the Bangkok waste-water system; improving a manual jacking system; making a device to prevent unbroken quartz/silica tubing from contacting the bottom of the collection pit; investigating the effect of traffic on a road tunnel ventilation system; developing an electronic compass; developing an alternative wax pattern die lubrication; providing automatic routeing of chutes for filling coal bunkers; and producing an audio guide for a Nature Centre for use by partially sighted visitors.

At the institutional level, the Royal Society has a Scientific Research in Schools programme which has been running since 1957. Initially it was meant to support individuals who were doing research for a higher degree but this has now changed to expect, indeed to require, the research projects to involve students themselves. The range of such projects has been enormous and their quality outstanding. Primary school pupils have done research into the natural stocking of a new pond; 14-year-old low-ability students have become authorities in establishing the conditions necessary for mallards' eggs to hatch; sixth-formers have become expert at detecting and analysing sun spots. A recent annual report recorded 22 schools working on such projects during the year including 'the palaeontology of key horizons in lower Cretaceous speeton clay in east Yorkshire', 'lead-free petrol and roadside pollution', 'sensing and control for the severely disabled', 'a study of some problems in extractive metallurgy', 'activity rhythms of the common oyster', 'ecological study of the greater horseshoe bat', 'hailstone collection efficiency of ice crystals and water droplets in clouds' and 'a 151 MHz interferometer for radio astronomy'.

The Institute of Physics has a Small Grants Scheme which is able to give financial support to teachers wishing to do projects with their students. Two examples will illustrate the scope of this work.

In the south of England, near the site of the Battle of Hastings, a teacher at a girls' school became interested in a sunken forest just off the coast. The sunken forest was well known, though exposed only at exceptionally low spring tides. What was not known, however, was when the forest had been submerged, when the sea had engulfed the coastline and moved it inland to its present position. The most common theory was that this stretch of coast had become overrun by the sea during the great storms of 1287. The girls had become interested in the age of the forest and wondered how they could find out. Had the Norman Conquest come through or sailed over the forest? Old maps did not give the answers, their history and geography teachers did not know, Kew Gardens and research groups looking into climatic change were consulted and provided background information but could not resolve the problem. They wondered whether carbon dating would be possible. Sure enough, after systematic analysis of the library and building on the science learnt in their A-level studies, they obtained their samples and, in cooperation with Harwell, obtained an authoritative dating for the forest of around 4,110 years, nearly 3,000 years earlier than previously believed. This simple, local question had lead the girls to experience the cross-curricular nature of many problems and gain experience of working with modern industry. They also experienced the involvement of the local media who shared their excitement in achieving a solution to a real local puzzle. In the words of their science teacher, Dr Fisher:

> Part of the romance of the project was that the pupils were walking on a stretch of coast line that played an important part in history . . . being the landing ground for the Norman invasion. That the pupils could use modern scientific techniques and reasoning to shed new light on part of our history certainly captured their imagination.

The TASTRAC project has also enabled school children to do scientific work of national and international importance. Initiated by a school physics teacher and a local university lecturer, it

Box 4.5   Two industry-supported student research projects

Project Superconductor

Having always prided myself as being one of those teachers who likes to think that I positively encourage any enthusiastic ideas put forward by my pupils I did not hesitate to say 'we'll have a go' when some sixth-form students said to me 'Can we make a superconductor?' Having only a very minimal knowledge of what a superconductor is, I have since become quite an expert following our little project. . . .

I began by reading the article in the *New Scientist* (30 July 1987) entitled 'Do it yourself superconductors', which gave a 'shake-and-bake' recipe for making one of these new high-T_c superconductors, most of the equipment being available in a typical American chemistry laboratory. Gary . . . a lower sixth science student and team leader of this project, then provided me with a shopping list of equipment required for him and his friends to make their own superconductor. The task seemed to be quite a difficult one. . .

The barium carbonate and copper oxide were readily available. However, finding the yttrium oxide was a little more difficult. Gary and I wrote to a local scientific company who replied almost immediately, giving plenty of advice and offering to order us some yttrium oxide . . . we were on our way!

We managed to charm the CDT department into allowing us to 'borrow' one of their kilns for the baking stage, but then had the problem of calibrating the kiln to suit our purposes. A chance conversation with a parent at the school's open evening produced a tremendous amount of valuable advice and also help in calibrating the kiln. The help given by friendly scientists from Culham Laboratory provided the sixth-formers with someone who was instantly accessible for on the spot advice.
The next problem we faced was how to press the superconductor after the first baking stage. It was back to the letter writing. Gary and I wrote to four different local industries, all of which replied offering their services. We finally arranged to use the press at Oxford University.

We were all set to start. The four sixth-form students involved carried out all the initial stages themselves.

At the end of the first baking stage, the headmaster was asked to switch off the kiln at the appointed hour, being the only member of staff around the school at nine in the evening. I then taxied the students and their superconductor to the University for the pressing stage. The final baking stage involved spending the night on the floor of my laboratory with alarm clocks ringing every two hours so that one of us could check the temperature of the kiln. (Not recommended for those teachers who like their home comforts, such as sleep and heating, especially if they have a full teaching load the next day!)

Once our superconductor was made, all we had to do was to test it. Obtaining liquid nitrogen from the local laboratories was the easy bit.

When tested, it worked.

We cooled the tablets by dropping them into a plastic beaker of liquid nitrogen, and then placed the tablet on the top of an upturned beaker. We found this was a better method than surrounding the entire beaker in liquid nitrogen and keeping the tablet dry. We had a sliver of a samarian cobalt magnet, about 2 mm, which we placed on the tablet.

To our delight, when using the tablet that had been heated twice, the magnet floated 1–2 mm above it. The magnet hovered with total stability and resisted being moved. The superconducting sections were obviously patchy, however. On one side no levitation could be observed and on the other the height of levitation varied.

Clearly then, the group had demonstrated that superconductors can be made in the school laboratory. It would have been possible to attempt the whole operation using none of the equipment (apart from the yttrium oxide and the magnet) supplied from outside school. However, knowing there would only be one shot at the manufacture we felt that doing so would reduce the chances of success unnecessarily. It was pleasing that the little 'experiment' with both tablets worked. From the students' point of view, it was a piece of genuine research.

The students concerned have the right to feel proud of their efforts. They found that a great deal of information given to them was inconsistent, and they had to make their own decisions at every stage of the project. The help given by all the local companies was invaluable, both in terms of the equipment given or loaned and in the time and effort of the people involved. The excitement of the 'research team' at the end of their successful project was proof enough for me to realise why it is that I find teaching a rewarding job.

How about having a go yourselves?

From artificial kidneys to novel catalysts

At a tertiary college, A-level students have been studying the chemistry of the cellulose dissolving properties of copper (II) complexes; this has led them into applications for dialysis membranes for artificial kidneys and for catalysis. These studies were supported under the Joint ASE–Royal Society Scientific Research in Schools Committee scheme.

As a young PhD student I once asked my supervisor, 'How does one get ideas for research projects?' 'It's easy. You simply go to a library, dust off an old tome on chemistry and flick through it until you find something that intrigues you or that you just do not believe. Go back to the laboratory and try it.'

Our research project on membranes for artificial kidneys and the subsequent work on novel catalyst and separation systems was born from just the approach suggested by my professor. Reading through an old chemistry book, I came across the fact that cellulose dissolves in cuprammonium, the beautiful deep blue solution that forms when copper (II) hydroxide dissolves in aqueous ammonia. This seemed a remarkable property. A group of second year A-level students were set the task of trying to find an alternative solvent for cellulose based on their A-level knowledge of transition metals and complex ions. The idea to replace ammonia as the complexing ligand with amines seemed obvious. Freshly precipitated copper (II) hydroxide was dissolved in aqueous methylamine and the resulting blue solution dissolved cellulose; success at the first attempt.

The foundations were laid for a research project

which has spanned many years and many A-level students. It has been a co-operative venture with students, staff and technicians working together. We have been fortunate in gaining considerable success, winning the BBC Young Scientists of the Year Competition twice, representing the UK in the Philips European Young Scientists and Inventors Competition, exhibiting at the Royal Society Conversazionnes (1983), having some thirty or so publications, and receiving a number of research grants. The success of research cannot and should not be measured by such material outcomes, however.

Far more important is the impact this type of work has on individual students. 'It was the most valuable part of my time at college' reflected one student during a visit to college during her summer vacation from university. Research has fostered a better mutual understanding of the needs of students and staff and has created an atmosphere in which students, not necessarily directly involved in the research, feel they are part of a college which is a vibrant and exciting place to learn. I have no doubt that project work has allowed students to demonstrate abilities that I would not have realised they possessed during my normal A-level teaching. Instead many have surprised themselves with what they have achieved.

Natural cellulose can be dissolved in a suitable solvent and regenerated in some desired form: fibres, films, hollow fibres. The viscose process dominates industry and has largely replaced the cuprammonium process of the last century. However, films and hollow fibres manufactured from cuprammonium are still considered to be the best material available for use in artificial kidney machines; there is a constant demand for improved membranes resulting in smaller, more easily used artificial kidneys operating at shorter dialysis times.

We were excited by our work, believing that we had contributed to an important area of applied scientific research, an area which had potential to improve the quality of life for thousands, as well as advancing our understanding of the systems at a more academic level.

was originally sponsored by the Institute of Physics small grants award but has now grown and incorporates financial support from UKAEA at Harwell. The project is based upon a plastic material manufactured at Bristol University under the name of TASTRAC, which is highly sensitive to nuclear radiation. It will readily record any alpha radiation passing through it. All that is needed to produce a national survey of the natural alpha radiation around the UK, or a measure of radon in domestic water supplies, even the spread of and retention of radioactivity around Chernobyl, is a supply of pieces of TASTRAC plastic, a yogurt pot, a coffee tin and some cling film, and an army of school children volunteers to collect the evidence systematically. Who needs gamma spectrometry and liquid scintillation counters, which accepted wisdom had previously deemed necessary?! Since 1987 two national surveys of radon emanating from the soil, two local surveys of radon in the home, in Somerset and in Cumbria, and a national survey of radon in drinking water have been conducted involving thousands of pupils in hundreds of schools. Again the excitement of being involved in front-line scientific research, producing significant data of national importance to the scientific and the public services communities, has been influential and effective in the scientific education of many school children. (The projects and results have been written up by Camplin *et al.*, 1988 and Allen *et al.*, 1993.)

Sponsored competitions

Most young people like competitions. They add that extra spice to the activity and are more fun. Though the activity must be satisfying and worthwhile for its own sake, an element of competition may add that extra spur required to get people started and the incentive to make that little bit more effort. Competition, as long as those competing believe that they have a chance of succeeding, can have a positive effect on motivation to produce one's best. More particularly, the element of competition and the receipt of external evaluation

and approval can do much for participants' self-confidence.

I remember doing projects with students in school, as part of their normal curriculum, and having good but rather self-deprecating work done by the students. When, however, they were persuaded to present these projects for public display on open days they were highly gratified to receive much interest and surprised approval from adults, many of them practising scientists themselves. They suddenly realised that their work was not 'just satisfactory for the teacher' but 'really good science in the eyes of the adult world of science and engineering', and that did wonders for their morale, motivation and self-confidence. When they refined them further for entry to a Young Scientist of the Year competition, sponsored by the BBC and the *Sunday Times*, the quality and professionalism of their work improved still further.

One particular incident occurred with students who had devised a method for recording photographically the different stages of the formation of a splash when a milk drop fell into a dish of milk. Many may have seen commercially produced photographs of the beautiful crown that develops, and the subsequent tiny drop that bounces back. The students had arranged a mechanical means of triggering the strobe flash, which could be systematically adjusted so that a photograph could be taken at a specified time in the development of the splash (see Box 4.6). A full film of still photographs was produced showing the stages in sequence. When these were displayed in public, for the competition, an adjudicating adult suggested that this was 'old hat', for such photographs had surely been taken before. With the confidence that comes only from complete ignorance, the student declared that in fact the professionals produced their photos by taking hundreds of photos at random and then rearranging them into sequence. The judge quickly made a few phone calls and confirmed that this was actually true, a fact that surprised him, myself and the student himself!

Many companies and professional institutions now sponsor annual competitions, with prizes and certificates awarded to the winners at local, regional and national level. The Royal Society of Chemistry had a Top of the Bench competition which consisted of a written test of basic chemistry, a research exercise on the theme of metals and a problem-solving exercise to make a timing device capable of measuring a period of 20 s from a test tube, water, rubber bands and Alka Seltzer. Duracell had a Schools Science and Technology competition to produce a useful invention, with prizes including £1,000 to the winners' school and an all-expenses-paid trip to Duracell World-wide Technology Center in Massachusetts, USA, for the winner. The Institute of Biology and the British Ecological Society had a Young Ecologist Award, a competition for 11–16-year-olds linked to BBC television's *Seeing through Science* programme open to students wishing to conduct an ecological study. The European Community has a high-status European Community Contest for Young Scientists for projects in any of the exact or natural sciences with an international jury looking for 'originality and creativity in the formulation of the basic problem, skill care and thoroughness in designing and performing the study, reasoning and clarity in the interpretation of results and quality of presentation'. Again, big prizes worth 5,000 ECU are awarded, plus fellowships to finance a placement in an industrial laboratory abroad. The Engineering Council has organised a Young Engineer for Britain competition for many years, with sponsorship from a variety of companies, and prizes in the most recent competition totalling £12,500. The Youth Section of the British Association, with British Gas, organises Masterminds competitions, with local and regional knock-out competitions targeted for teams of 11–18-year-olds from BAYS (see Box 5.4 in Chapter 5). The British Association, with the ASE and Nuclear Electric, also organises an educational competition for the 9–13 age group called The Science Challenge emphasising communicating scientific principles to lay people, consisting of school-based work, a project at the finals and an interactive quiz. The *Times Educational Supplement*, supported by BP and CREST, has sponsored a TES Environment Award, organised by CREST, which 'aims to encourage teachers and their students, in co-operation with local business, to undertake fulfilling science and technology project work about environmental issues'. The Chemistry Club had a Challenge for '11–14 year old pupils to design and demonstrate practical chemistry experiments to illustrate themes or topics'.

Perhaps a brief report on the National Final of one of these competitions, the TES Environment Award, will give something of the flavour and value of such competitions. About 2,000 schools had entered for the award, and each of them would have benefited from doing its project. From involvement they, and often the whole school, would have become more aware of environmental issues. Thirty-two schools from across Britain, with pupils aged 10–19, had been selected for the final and brought their results and displays to a gallery in the Natural History Museum in London, itself a prestigious and stimulating location for the students. On the day of the final this gallery was filled with the 32 displays, each presented in a lively, graphic way with photos, charts, graphs and models. Alongside each, and most impressive of all, were the students who had been responsible for developing the projects. The enthusiasm and confidence with which they talked about their projects demonstrated clearly the depth of their involvement and the relevance and importance to them of the work they had done. The groups came from a variety of schools and abilities, from first-year pupils in a comprehensive school to A-level candidates in sixth-form colleges, from girls in selective grammar school to children in special needs schools. All had worked on an aspect of the same problem, to improve the environment. All shared the same enthusiasm and pride in their achievement.

Projects included:

- developing a garden suitable for people in wheelchairs;
- investigating ways of saving energy in school (which actually saved their school £7,000 in the year);

Box 4.6   *Controlled high speed photography of milk splashes*

In a lower sixth A-level physics set, a group of boys decided to investigate how splashes were formed. Originally the aim was to emulate, and if possible improve upon, some of the photos of splashes seen in various textbooks. It seemed that work in this field had previously been done in one of two ways: either by using a very high speed, and very expensive, cine camera, or by taking hundreds of still photos at random with a flash gun and selecting the appropriate picture. The group decided to try and control the instant when the photo was taken, so that each successive stage in the formation of the splash could be studied. The principle they developed was in essence beautifully simple, and after a considerable amount of experimentation with various prototypes, has developed into a very reliable and versatile piece of apparatus.

The splash being studied is formed by a drop of salt water falling from a burette into a dish of milky water. The drop falls between two wires, making an electrical contact in a relay circuit as it does so. This triggers the release of a steel sphere from an electromagnet. Thus the sphere and the drop fall side by side. Beneath the electromagnet two contact wires are mounted so that the sphere falls on to, and through, them. This makes an electrical contact in the external circuit of a Xenon flash unit. Thus if the experiment is done in a darkened room with an open shuttered camera, the flash will enable a picture of the splash to be photographed (see Figure 1). By extending the distance between the electromagnet and the lower contacts, the time taken for the steel sphere to fall is increased, and the time into the life of the splash at which the photograph is taken can be increased. In this way a sequence of photographs can be taken which reveal the different stages in the formation and decay of the splash – and a fascinating life it is too! The time into the development of the splash at which each photograph was taken can be calculated from the height through which the ball falls, or by using an electric millisecond timer triggered by the apparatus itself. In both ways an accuracy of a millisecond can be established. Repeating the photograph of the splash for the

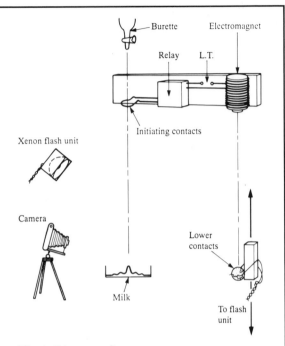

Fig. 1 *Diagram of apparatus*

same setting of the apparatus confirms the consistency and the reliability of the method.

The stages in the life of a splash vary to a slight extent on such factors as the size, shape and speed of the incident drop, and the depth and physical properties of the liquid. But the following sequence is typical: a drop entering the liquid quickly forms a crown which grows and then decays back into the liquid forming a crater at the centre. From this crater a column grows which reaches a maximum and starts falling back. As it drops back it splits up into an upper spherical drop almost suspended above a shortened skittle-shaped column. After these have fallen back into the parent liquid a smaller secondary column emerges which itself splits into a tiny drop and shortened column. As the central series of columns rise and fall a series of waves radiate from the point of impact of the drop. This sequence was obtained from a drop travelling at 300 cm s^{-1} into milky water 3.5 cm deep. The time from entry until the decay of the second column is about 0.3 seconds.

This apparatus can be modified to take

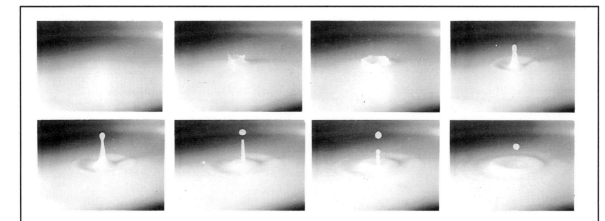

photographs of other systems at a specified instant of time. It has been used to study the effect of a steel ball-bearing falling on to a light bulb and also on to a sheet of glass. For this two electromagnets are used, mounted side by side, and a steel ball suspended from each. When the magnets are switched off the two balls fall side by side, one on to the light bulb, say, and the other on to the trigger contacts.

The boys responsible for the design, construction and execution of this project took a problem which, though readily comprehensible, provided some very searching conditions. By applying basic principles, and using ingenuity and originality, a design was evolved which, after various less satisfactory attempts, proved adequate and thoroughly reliable to solve the original problem. And this is what real science is all about. What is more, the apparatus proved versatile and cheap, being made literally from pieces of wire and bent pins; apart from standard laboratory apparatus the cost of the project was the price of a bottle of milk!

- developing a ramp for frogs to get out of a recently paved pool;
- designing and making a system to enable used bathwater to be redirected from the drain to the garden and water the plants directly;
- investigating links between the aluminium content of local water supplies and the prevalence of Alzheimer's disease;
- and developing and using a method to investigate the biological indicators in sludge from the local sewage treatment plant.

The judges, scientists and industrialists from the sponsoring organisations found the task of discriminating very difficult and the quality of the work surprisingly high. 'I am beginning to think that all the press reports of low standards of science in our schools are a lot of hogwash', said one eminent industrialist. Eventually winners were selected in different groups, but none of the schools felt that they had lost: all had achieved much through participating and seeing their work so widely appreciated by adults and their peers.

Two specific aspects struck me, apart from the variety, quality and obvious value of the projects to both the students involved and their community. The first was the number of girls involved in the projects: just over half of the finalists were girls, and this at a time when it is said that girls do not like school science. They obviously enjoy this type of science and do very well at it. The second was the range of students tackling a similar project, each in a different way and each to different depths, but each entirely satisfying at its own level. Because the problem was open-ended and tackled to the limits of the abilities of the students involved, it gave them all challenge and fulfilment at the appropriate level. This was 'differentiation by outcome' from a common open-ended task. The National Curriculum, with its artificial structure of ten levels of achievement, imposes a quite unnecessary and harmful form of differentiation on schools. Given the opportunity to tackle real science problems, such discouraging differentiation is not needed. The day, and the competition itself, was hard work, demanding and tiring for all concerned. But it was also fun, stimulating and highly satisfying for the students, teachers, judges and visitors alike, and showed what effective science teaching in schools can produce.

The plethora of such competitions organised and funded by different organisations (and those listed above are only a sample), illustrates the value both in learning science and in motivation which many believe such competitions engender. Anyone who has been personally involved, either as a student, a teacher or, often to their own surprise, a scientist or industrialist who has come in to judge, will know that this is true.

Personal hobbies

Perhaps we should not let this chapter end without some reference to the student research projects which go on out of school for many young people, but which so often go unrecognised and unutilised in school science. Many students, often encouraged by parents or teachers, will spend many hours looking after their pets, developing a garden, breeding pigeons, studying the stars and the night sky, analysing and recording the weather conditions, fishing, building electronic devices, model-making, working with local environmental or archaeological groups, making jewellery, mending clocks or motorbikes, making music, even looking after the home or baby brothers or sisters. Many of these hobbies, which engross and inspire the students for many hours over long periods of time, are highly science-based and enable students to personalise their scientific understanding and become expert scientists in those fields. Many sports and athletic activities have an increasingly scientific foundation, too. Often students' knowledge will be 'tacit' rather than 'explicit', but it is no less important for that. Real scientists rely on tacit knowledge quite as much, if not more, than on that knowledge that they have formally gained and can make explicit. The trick for the science teacher is to bring that knowledge and expertise into the class room, to build on it and link it to the science in the school curriculum, and to make constructive links in the pupils' understanding between the two domains of science – everyday science and scientists' science. Not only will it develop students' scientific understanding but in celebrating students' own knowledge, their self-confidence, self-esteem and motivation will be increased. Such hobbies, too, can be harnessed to make school science teaching more effective.

CHAPTER 5

Stimulus science activities

Many scientists, many science teachers, when recollecting the influential events that led to their liking for, and commitment to, science will point to one or two events that stimulated their imagination: meeting an exciting real scientist, perhaps, attending a spectacular lecture demonstration or being shown the wonders of the stars, the fossils in a quarry, the life in a rock pool. The role model of an inspirational science teacher has been influential for many.

The influence of a scientific home, with the absorption of positive attitudes towards science and the encouragement to 'fiddle with mechanical things', to look after animals, to study and appreciate the environment, certainly has encouraged many to follow the same route. Roe (1953) in his study of successful, creative scientists, concluded that the home background was the formative factor in their career choice, with the habit of learning for its own sake being particularly significant. Subsequent work by MacKinnon (1962) concluded that the home factor that was influential in the career choice of creative architects was 'an extraordinary respect for the child and confidence in his ability to do what was appropriate . . . [which] contributed immensely to the latter's sense of personal autonomy which was to develop to such a marked degree'. Parents often exert influence not through overt advice towards a particular career (Breakwell et al., 1988), but through the attitudes and interests which the child subconsciously absorbs. The shared 'Protestant work ethic' and belief in the integrity of knowledge can also be 'absorbed with the mother's milk'. The influence of home background was illustrated in my FASSIPES research in which the proportion of science sixth-formers having a parent with a degree in science or engineering was almost twice that for those not heading towards science (45 per cent compared with 23 per cent). For those heading towards a degree course in physics the proportion was particularly high (53 per cent). A potential engineer said: 'I think it's my father's influence. He's an engineer and I've helped him around the house with a lot of things . . .' A potential chemist: 'I think I was guided into the science area by my dad. The fact that my dad was a chemist influenced me a lot.' Fathers were reported as being considerably more influential than mothers.

The influence of class, with the split between 'the cultured amateurs who rule us and the dirty-handed scientists who have the idea and do the work', runs deep in the English culture (Wiener, 1981) and has influenced career choice away from the sciences for many members of the 'upper classes'. In his study of the background of scientists Galloway (1991) concluded that 'scientists and technologists often have working class backgrounds and that science is not perceived as a suitable career by better off or middle class parents'. In the words of a head of science at an independent fee-paying school explaining why he felt more students did not go into careers in science: 'Having paid vast fees they (the parents) expect a well-paid job in return.' The proportion of English scientists who came from the old grammar schools, particularly the old Northern grammar schools, is very high.

It is obviously not possible to change a student's parentage, school type, or home background (nor easy to change the prejudices of society!), but we can transfer something of the 'respect for the child and confidence in their ability to do what was appropriate' which MacKinnon found so important and provide something in the way of stimulus activity that will be picked up instinctively in a scientific home. Furthermore, we can provide extra stimulus activities for students which might well fire their imaginations and influence their attitudes to science and engineering.

Coming from a non-scientific home background and having a competent but uninspiring science education at my (grammar!) school, I can still remember the formative influence of three things. First, a beautiful stoboscopically lit wave machine reaching right from the floor to the ceiling of the Royal Institution at one of its children's lectures (by Sir George Thomson, I think, though it was the fascination and wonder of that wave, not the fame of the speaker, that I remember). Second, a project that I was given at the end of my sixth-form course to make a Geiger counter tube from basic materials – it was the first time in my school career that any teacher had shown faith in my ability to work on my own initiative (although the tube never really worked efficiently, the boost to my self-confidence was considerable). And third, sitting at the back of a packed lecture hall and hearing Niels Bohr speak, in not very good English, about the development of the atomic theory, as if he himself had had little to do with developing that theory. What all of these had in common, and why they were so influential, I suspect, was that they appealed to the affective rather than the cognitive part of my nature. They stimulated my imagination and thus triggered and strengthened my commitment to science, in my case physics.

In this chapter I will share some of the ways teachers and scientists have stirred the imagination of their students by providing stimulus activities in science.

Lecture demonstrations

The Royal Institution lectures

Undoubtedly, the most prestigious and influential lecture demonstrations have taken place at the Royal Institution (RI) in London. Founded in 1799 by the extraordinary and romantic adventurer Count Rumford to 'disseminate and promote the rapidly developing sciences to the public at large', it has proved a centre for scientific discovery and discourse ever since. In 1826 Michael Faraday, then resident professor at the RI, instituted the Friday Evening discourses (for scientific, political and social leaders) and the Christmas Lectures (attended by their children) to 'diffuse knowledge and facilitate the introduction of inventions by popular philosophical lectures'. These have continued through to the present day. Given in the splendid semi-circular lecture theatre, built in 1801, packed with 500 children and teenagers, the leading scientists of the day present their lectures in comprehensible language and with a multitude of spectacular demonstrations which rarely fail to stir the intellect, imagination and involvement of the children. Michael Faraday, as brilliant a lecturer as he was scientist, started the tradition with lectures on his work on electromagnetism. Many great scientists were invited to give subsequent lectures: John Tyndall, James Dewar, William and Lawrence Bragg, George Porter, Charles Taylor, Eric Rogers, and Richard Dawkins, to mention just a few. Since the early 1970s the Christmas Lectures have been televised, thus extending their influence to millions of young people unable to attend the lectures themselves. In addition, the RI offers a programme of lectures throughout the year for school children within reach of London, as well as exhibitions illustrating the work of many of the famous scientists who have worked there. I suspect that many, not only myself, have been awed and inspired by the sight of Faraday's early apparatus and laboratory, and encouraged by the almost unintelligible writing in his notebooks!

Fig. 5.1 The Royal Institution Christmas Lecture of 1961, given by Sir Lawrence Bragg, from an oil painting by Terrence Cuneo

Professional institutions' lectures

More recently, some of the scientific institutions, such as the Institution of Electrical Engineers (Faraday Lectures), the Institute of Physics, and the Royal Society of Chemistry, have instituted lecture demonstrations for schools by leading scientists and science populisers. These have been taken around the country and presented to large numbers of young people, introducing them to a quality of scientific demonstrations more spectacular and more impressive than would be possible by science teachers in their own laboratories.

The popularity and range of these lectures can be illustrated by the Institute of Physics Lectures for Schools. These were started in 1983–4 with Cyril Isenberg lecturing on 'The Detection of Earthquakes and Underground Explosions' at six venues around the country. It is estimated that over 10,000 children heard Mervyn Black's lecture, 'Say it with Frozen Flowers', in 33 centres. The size and scale of these excellent ventures is formidable, and made possible only by the support of the lecturers' employers and by finance from industrial sponsors.

Local BAYS clubs

There can be very few areas in Britain where there are no scientists or engineers able and keen to come and talk to students about their work. Care needs to be taken in selecting and preparing such experts for their student audience: a boring or incomprehensible lecture can do more harm than good and reinforce all of the worst stereotypes about dull scientists. It is useful to invite a new lecturer to come to the school and talk with some of the students who are to be the audience some time in advance of the lecture so that he/she can tune the level and content of the talk appropriately (or withdraw if it becomes clear that communication is going to be a problem!). It is, of course, good if the speaker can spend more time in the school and engage with the students in 'normal' class activities, too, so that there is an opportunity to build up personal relationships and for the local scientist to become part of the school science team.

He or she might become a useful adviser or contact for the students' research projects. Lists of recommended speakers are available from most of the national institutions for the different sciences, and the Neighbourhood Engineer scheme organised by the Engineering Council is also a useful source of speakers. Other speakers are readily available from the science departments of local universities, who often produce their own programme of speakers and open days, or from hospitals, or from local science-based industry.

Many schools, or in some areas groups of schools, find that the establishment of British Association Youth Section (BAYS) clubs provides an excellent structure for bringing good scientists into the schools as speakers. BAYS is a national network of science clubs aimed at students between the ages of 8 and 18. It is the youth section of the British Association for the Advancement of Science, a national organisation promoting science and technology. BAYS membership involves

---

Box 5.1   A programme of BAYS lectures

1991 AUTUMN TERM	Regional Events	1992 SPRING TERM
Tuesday, September 10th (Sherborne) 'LEUKEIMIA – CAUSES & TREATMENT' Dr. S. ROATH University Dept. of Haemotology Southampton General Hospital	**CHRISTMAS LECTURES** 16th, 17th, 18th December 1991	*Tuesday, January 21st (Sherborne)* 'THE IMPORTANCE OF SENSES IN OUR LIVES' Dr H. V. WHEAL Dept. of Physiology and Pharmacology University of Southampton
Tuesday, October 15th (Yeovil) 'FORENSIC SCIENCE' Dr M. W. DUCKWORTH Home Office Forensic Laboratory Chepstow	**IDEAS INTO ACTION** and **MASTERMIND REGIONAL FINALS** 8th February 1992	*Tuesday, February 11th (Yeovil)* 'LASERS AND THEIR APPLICATIONS' Dr P. APLIN Dept. of Physics University of Bristol
Tuesday, November 12th (Sherborne) 'FOOD ADDITIVES AND BEHAVIOUR' Mrs M. Pinder Dietician Dorset County Hospital Dorchester	**SUMMER LECTURES** July 1992	*Tuesday, March 3rd (Sherborne)* 'CRYOGENICS' Professor R. G. Scurlock Institute of Cryogenics University of Southampton
Tuesday, December 10th (Yeovil) 'ENVIRONMENTAL ISSUES' Mr I. MAXWELL Environmental Health Officer South Somerset District Council		*Tuesday, March 24th (Yeovil)* 'EXPLORING MUSIC' Professor C. Taylor

PLEASE NOTE: LECTURES ALTERNATE BETWEEN YEOVIL COLLEGE AND SHERBORNE SCHOOL FOR GIRLS.
ALL LECTURES COMMENCE AT 7.00 pm. Teachers and Parents welcome

Fig. 5.2 Optics activities in a science club:
(a) mounting a camera to the school telescope for the
     stellar photographs; and
(b) a photograph taken with a pin-hole camera

more than just scientists coming to give talks or
lecture demonstrations; it also involves club activities, competitions, quizzes, CREST awards and
BAYSDAYs, which are described below. Box 5.1
illustrates a typical range of talks for a BAYS
group in the South of England.

Science teachers' demonstrations: a dying art?

Even more local than the local scientist as a source
of lecture demonstrations are the school's science
teachers themselves. The lecture demonstration
used to be – and in many countries still is – at the
centre of school science teaching, but it has now
been replaced in many schools in England and
Wales by the ubiquitous class experiment. I would
like to see a revival of the dying art of demonstrations, with teachers giving some priority to the
impressive, large-scale theatrical presentation of
science. Few have the time or the resources to
produce full-length demonstrations, but the

impact and influence of good demonstrations in
the course of a lesson cannot be too highly
emphasised. Apart from the cognitive benefits,
with the teacher talking through a scientific principle around the demonstration experiment to
focus students' attention, the affective benefits of a
memorable demonstration can be considerable.
And, as we have argued before, it is the affective
area in which students' imagination and interest
are developed that is all-important for long-term
commitment.

Sources for good demonstrations are few, and
remain the product and the possession of individual science teachers' own creative minds. The
teachers' notes in the ASE's journal *School Science Review* contain a wealth of good ideas. I
would also recommend Charles Taylor's (1988)
book on *The Art and Science of Lecture Demonstration*, and the new background book from the
Royal Society of Chemistry, *Chemical Demonstrations for Schools* (Lister, 1994).

Of course, the best demonstrations are often provided *in situ*. The biologists and geologists on their fieldwork, the physicists looking at the stars or a local building site, the chemists taking their students to the local brewery or industrial chemistry plant, provide students with experiences well worth the organisational effort required.

Hands-on science centres

Traditional science museums contain collections of, usually old, scientific and technological objects displayed, often enclosed in glass cases, with concise notes explaining their origins and significance. The visitor is expected to walk around, to look, to absorb the information and to move on. The visitor's role is passive, there is no opportunity to interact with the exhibits: 'Don't touch!' signs are not uncommon. The aim is to inform the mind. The hope is that the visitor will be awed by the range and totality of the exhibits.

Many of us will remember visits to such exhibitions in our youth. There was, of course, the excitement of a day out, a visit to London with our parents, perhaps, or in a school party with our friends and packed lunches. But I suspect that little was remembered of the actual science, little mental illumination gained, and that the sheer weight of all the exhibits left us bored rather than awed, tired rather than inspired.

Looking back over forty years or so, I can remember my visits to the Science Museum in London, with its vast collection of technological objects in layer upon layer of endless galleries. But it was not the enormous steam engines which filled the central hall with great, shiny, cold flywheels and pistons that held my attention, or the early flying machines or even the vast, strange, slowly swinging pendulum suspended from the ceiling some six floors above. It was the children's gallery in the basement that drew me, fired my imagination and left me with memories which are graphic to me still. For it was there that I could be involved, could interact with the exhibits and relate to the ideas personally. There were still

some objects in glass cases, but most of those had buttons to push or handles to turn which moved parts of engines or illuminated different aspects of house interiors. But the most memorable exhibits involved me actively and left me with puzzles rather than answers. There was the glass-sided passage through which you walked and the door ahead opened apparently of its own accord, as you walked through a triggering light beam. There was the collection of large pulleys and weights, which allowed you to pull the handles and lift the heavy sacks easily or with difficulty depending on the pulley arrangement. But most wonderful of all was the unattainable 'gold' ball, set in the middle of a circular table, which disappeared down into the table every time you reached over to grab it. Nobody explained how it worked (perhaps your arm changed the inductance of a loop of wire set in the table top) but the fascination of the puzzle, the challenge of the problem which was so nearly attainable, and the sheer fun of the task drew me, and may other youngsters, to it and left me with a sense of wonder.

These were the really influential exhibits, because they caught my imagination and forced me to be active and ask my own questions, rather than be passive and receive the given knowledge.

The children's gallery of the Science Museum was the precursor of the hands-on science centres of today. During the 1980s interactive science museums were established in most countries and have proved enormously popular with both children and adults. The first truly hands-on science centre was the Exploratorium opened in San Francisco in 1969, which was followed by another in Toronto. In England most of the big centres of population now have one, ranging from Launch Pad in the Science Museum in London to the Exploratory in the Old Station at Bristol; from an exploration of vast universes at the Centre at Jodrell Bank Telescope in Macclesfield to the examination of tiny creatures at the Micrarium in Buxton. Each has its own distinctive flavour but all have certain features in common. They are interactive, with specially constructed large, attractive exhibits that encourage visitors to investigate

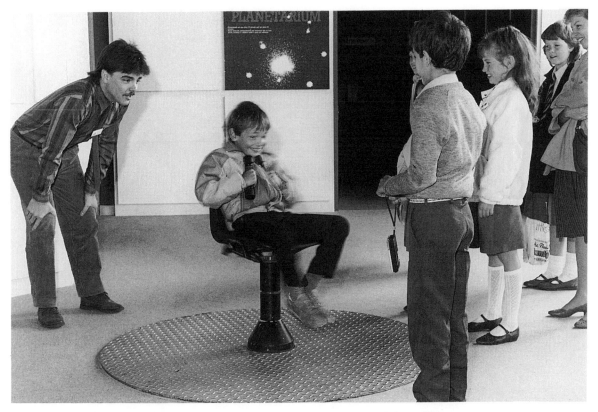

Fig. 5.3 Experiencing the delights of circular motion at the Jodrell Bank Science Centre

natural phenomena and experiment with technology. The exhibits are contemporary rather than historic, striking and dramatic in their impact. They are informal places – with young 'explainers', 'guides' or 'pilots' on hand to welcome, to discuss the exhibits and help if required (but in the role of friendly stimulators of enquiry rather than authoritative teachers with the right answers). They are orientated to the general public as well as educational groups, aiming in different degrees to enlighten as well as entertain, to stimulate the mind as well as excite the emotions. Finally, they are fun places, aiming to make learning enjoyable and memorable. There is only one way really to appreciate the value of such centres. Go to one. Get involved with the exhibits, watch the children's involvement and talk with them.

There are, of course, ways of killing the effectiveness of a visit to such a centre. Probably the best is to take it too seriously and treat it as a vehicle to teach the youngsters some predetermined science. Periodically, one finds a group of school children burdened with worksheets, going round the exhibits, seeking answers to their teacher's questions. Such direction kills the children's own curiosity and sense of wonder and prevents them from testing themselves on their own questions. Adults, and especially teachers, want to know what the children have learnt from their visit. But it is not the acquisition of knowledge that is the main purpose of such visits. They will acquire some new scientific knowledge, different children will learn different things, but much of that learning will be tacit as they gain personal experience of different phenomena. Over and above the cognitive gains will be the affective ones, for the students will have had their curiosity aroused, their imaginations fired and their motivation

Fig. 5.4 Wondering and speculating about the properties of black holes at the Jodrell Bank Science Centre

strengthened – often making a deep impression that will prove a permanent foundation for future learning and career aspirations.

If you are still unconvinced about the merits of such centres, and you really are undecided about making a visit, I would recommend Jerry Wellington's video on *Hands-on Science Centres* (available from the School of Education, Sheffield University) and the September 1990 edition of *Physics Education* (from the Institute of Physics, Bristol) which has a series of articles on hands-on science. But, as with the centres themselves, there is no substitute for personal, active involvement.

School–industry links

School science teaching can become limited by the school environment, using only the resources that science teachers can bring through their own

expertise and teaching skills and through the laboratory equipment available in the school laboratory. To overcome this limitation teachers have increasingly been looking to widen their students' experience by utilising the resources of local industry. The effect of this, when done well, has been to increase students' appreciation and understanding of science as it is seen to be applied in real situations, with real equipment, and with real engineers and scientists. Both the content and the processes of science become real, in all its messy excitement, as it is set into its social, economic and technological context.

There is no single pattern for the ways schools and industry can work together, because each relationship will depend on the local situation, and in particular on the personalities, interests, energy and expertise of the personnel involved. The key idea is that of partnership, with school and industry working together to share experiences to the

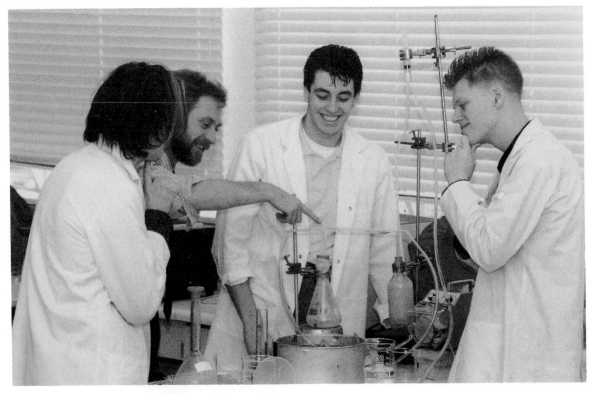

Fig. 5.5   An industrial adviser helping with a biofermentation project

benefit of the students. No amount of central organisation will make such schemes work if the personnel involved do not share a mutual respect for each other's enterprise. Although certain things can be organised centrally, and some of the bigger companies like ICI have a splendidly supported network of industrial liaison workers, the practicalities of how, and whether, such links will develop successfully have to be worked out locally over a period of time.

Typical ways local industry can support schools might include:

- providing speakers, though these need to be selected and prepared thoroughly if they are to relate positively to the students and to their science curriculum (bad speakers can do much damage to the image of science and of industry);
- enabling visits for students, though again these need to be prepared for by a preliminary visit by the teacher, who can then advise on a suitable programme and prepare the students appropriately;
- providing consultants who can advise students on their projects; providing ideas, problems and resources for students' projects;
- providing ideas, problems and resources for students' projects;
- helping produce enriched curriculum resource material;
- supplying support, judges and prizes for technology days, invention competitions or great egg races;
- providing advisers for CREST award projects;
- organising student in-house work experience,
- providing opportunities for teachers to gain experience of work in industry either in a short-term (one- or two-week) Teacher Placement Scheme, or in a longer (one-term or one-year) secondment;

- supporting Young Engineers Clubs in the school;
- providing advisers for Young Enterprise Schemes;
- producing careers advice, Careers Conventions and mock interviews.

To foster such constructive links a Neighbourhood Engineers scheme has been developed by the Engineering Council. This was launched in 1987, with DTI funding, with the aim of having four practising engineers linked with every secondary school in the UK. The scheme is organised in 19 regions around the country and already most schools in England and Wales are involved, with over 20,000 engineers giving some of their time to help improve science and engineering teaching in schools. The Engineering Council has produced a most helpful publication (Bridges, 1992), which will benefit anyone involved in developing such work. It gives both practical guidelines and exemplars illustrating Neighbourhood Engineers in action.

The spirit of Neighbourhood Engineers is well summed up in the words from their leaflet *Practical Support for Schools* (Engineering Council, 1992):

Neighbourhood Engineers provide friendly, informal, practical and committed support to the daily life of their local schools. Working closely with teachers, the engineers form effective teams to consider how best their ingenuity and resourcefulness can develop the 'Engineering Dimension' within their local school. Based on the concept of partnership, teams of teachers and engineers work together to meet the objectives set for the school. This team effort assists with the delivery of the curriculum and enhances the economic and industrial understanding of young students. Neighbourhood Engineers help with project work in the classroom, providing their school with examples of up-to-date technology. The value of the scheme lies in the long term, committed relationship between teachers and local engineers. The engineers become 'friends of the school' personally involved in the life of the school and the needs of young people'.

Where this is working well – and, despite the practical logistical difficulties involved, most of the schemes are working well – teachers receive practical and moral support for their work and students receive inspiration and guidance of a very high quality. Many students have gained their first insight into the fascination of science and engineering as careers through working alongside an enthusiastic and expert neighbourhood engineer (see Box 5.2).

Fairs and fun, facts and fantasy

On top of all the exciting science taught in schools, the science research projects students take up and the more regular sponsored activities described above, there is a variety of other activities which can be highly stimulating to a young potential scientist.

'Science at Work' or 'Physics at Work' exhibitions, sponsored by the Institute of Physics, enable students to meet a variety of practising scientists and talk with them about some of their latest exciting scientific projects and also about the work of a scientist. Careers Conventions also provide an easy way for young people to find out something about the potential for a career in science and engineering.

Chemistry Club, sponsored by Salters Institute of Industrial Chemistry, is helping to revive the habit of chemistry, and science, clubs in schools. Science clubs, photographic clubs, electronics clubs, radio clubs, animal clubs, hobbies clubs, engineering clubs, astronomy clubs, gardening even young farmer's clubs, have all done honourable service in the past and provided the opportunity for many students to develop and fulfil their scientific interest. Many future scientists have become stimulated and confirmed in their scientific interest by involvement in such school clubs. Unfortunately, such activities in many schools have been forced out by the pressure of teaching and implementing the main school science curriculum. Such clubs do demand time and imagination from the supporting science teachers. We must

Box 5.2   Neighbourhood engineers in action

*The following account describes the way in which the
Neighbourhood Engineers scheme has worked in
comprehensive schools in Devon and Cornwall. This
example is not selected as an exceptional case, simply
as an instance, of the scheme in operation in
circumstances where it has had a reasonable chance
to get established.*

*It is an abridged version of a case study which
appeared in the evaluation of the scheme produced by
a team from the University of East Anglia in the
summer of 1991. A fuller account of this and other
case studies appear in the report available from the
Engineering Council (1992).*

FOUR SCHOOLS IN DEVON AND CORNWALL

Of four schools in this study, three were situated on
the edge of small, attractive, rural towns, and the
fourth situated in the centre of a small rural market
town. Although only one of the schools is a
Community College, there is a strong community
dimension in all of them. Conscious of their
geographical positions, all four schools have
initiated and developed innovative approaches to
the enrichment of their curriculum including strong
'school–industry' links.

Making the links   Once the Regional Manager has
recruited the required number of engineers for a
school (usually five), the headteacher and the
Manager organise an inaugural panel meeting at the
school in order to begin the partnership. The
elements of the meeting usually consist of:

- the Regional Manager setting the scene by
  reviewing good practice;
- local engineers and teachers who are already
  working together in another school describing
  their experiences;
- an informal buffet supper, with an opportunity to
  establish new relationships;
- making arrangements for the first meeting of the
  school panel; and
- an opportunity to raise questions.

People attending this meeting usually include the
Regional Manager, visitors from an experienced

partnership, the new neighbourhood engineers, the
headteacher, teachers who have either chosen to
come or have been invited to the meeting, and
sometimes members of the school governing body.

After this the new individual partnerships or
'panels' are free to work in any kind of collaboration
they find most useful. Regular group panel meetings
are held in order to network and facilitate an
exchange of progress and ideas between active
panels. The Regional Manager continues his links
with all schools and engineers by participating
informally in some of the activities and by organising
and attending many of the group panel meetings.
Also, he is always available to schools, teachers and
engineers in order to facilitate the smooth running
of the scheme.

What are the best uses that have been found for engineers?

'There is no set pattern. Engineers are all individual
personalities. They all have their own expertise and
specialisms, so you do what you can.'
(Neighbourhood engineer)

The nature of involvement between a school and
its engineers is dependent on a number of
interrelated factors. On the face of it, a major factor
is concerned with the balance of power in the
partnership. In some schools, the school initiates
and drives the activities that the engineers become
involved in; they are there to service the needs of
the school and the pupils by offering time, expertise
or resources to a school project or a curriculum area:
'The school initiates and leads because we know
what our needs are, and what we can handle.'
(Headteacher)

In other schools, the partnership may have begun
on this basis, but the relationship has changed over a
period of time to one of shared initiatives: 'Their
first moves were coming to and helping with things
that we organised. They gained confidence dealing
with the kids. They came to know the staff, and then
they actually said "There's something we really want
to do." It was great.' (Teacher) In this example, the
engineers worked with the teachers on the planning,
execution and evaluation of the project as equal
partners.

Box 5.2   Neighbourhood engineers in action (continued)

Activities that Neighbourhood Engineers have participated in or initiated

- *Exhibitions:* A full day, 'hands-on' exhibition 'Engineering for People' at a school involving stands from a large number of engineering firms (of greatly varying disciplines). This included a fair number of female engineers. Pupils from four year groups spent time at the exhibition, which was considered very successful by adults and pupils alike: 'It was a great success – buzzing all day . . . what we wanted to do was to make pupils realise that engineering is not just about people wearing overalls and working with a greasy spanner.' (Teacher). This exhibition, which was initiated by the engineers in one school, worked so well that it is being repeated in other schools.
- *Careers:* Engineers come in and talk about engineering to pupils during their careers work. They also provide materials on careers in engineering. Some engineers also hold 'mock job interviews' with the pupils.
- *Classroom work:* Some engineers work with pupils in the classroom by giving inputs to specific topics, for example, 'flight' or 'energy'. One engineer works in the classroom (in collaboration with the teacher) on training in technical and vocational activities with pupils who have learning difficulties and other special needs.
- *Resources:* Help with outside resourcing. For example, a pupil needed to use a certain metal process that was too expensive to do in school. An engineer took the object away and had the process completed in his company. It 'made a great difference to the quality of his project' (Teacher). Another engineer (through his company) gave the school some engineering equipment it couldn't afford.
- *Engineering clubs:* Engineers help with Engineering Clubs (after school) by offering input and working on projects with pupils.
- *Lectures and talks:* In one school there is a regular after-school talk on aspects of engineering given by the engineers. Usually about 30 pupils attend (this includes a few girls).
- *Extension classes:* A number of engineers come to the IT and Electronics after-school extension classes in one school and work with the pupils and teachers.
- *Working on sites:* Pupils and teachers go out on to sites with engineers to work with their companies and industries. An example of this was when a group of pupils were invited to observe and research (over a number of visits) a working site for minor problems. When some had been identified, the pupils took them on as problem-solving projects in the school: 'This helps us and it helps the firm.' (Headteacher)

  An engineer in another school took a group of pupils (with the teacher) on a real job he had been commissioned to do, because it fitted in with their school project on energy. With the help of the engineer, the pupils did the tests, analysed the data, and wrote up the final report.
- *Project work:* Teachers and engineers collaborate on setting up school or class projects, some of which the pupils use for their GCSE work: 'In our school there are three of us [engineers] who are interested in heating and ventilation, so we have got an energy project on the go.' (Neighbourhood Engineer) This project has now also spread to the feeder (primary) schools.
- *School visits:* Engineers initiate or help (and sometimes fund) school visits to places and exhibitions which show different aspects of engineering (for example, a visit was arranged to an office in British Telecom). The pupils expressed surprise at what they saw; mainly female engineers, working in an office: 'I didn't expect to see people in an office. I thought they worked outside, mending things . . . I suppose that's why we think engineers are men, because they do manual things.' (15-year-old female pupil)
- *Participation in other activities:* Engineers become involved in other, larger activities. An example was a Maths and Industry Day (where a number of problem-solving activities for pupils were set up). A number of local and regional industries sent some of their people along and this included a group of engineers (men and women) from one firm.

- *Equal opportunities:* Engineers collaborate with schools in order to initiate activities which promote positive images of women in engineering careers and occupations.

There are many other, less formal, involvements. As the engineers become more embedded into school life and build up strong relationships with teachers, the links become more flexible and adaptable to individual needs.

What do staff think they gain by working with engineers? Teachers pointed to four main benefits of working with engineers:

- Access to the new and interesting knowledge, skills and resources that the engineers were able to offer teachers. This was not only considered to be good and effective in-service, but also fascinating in its own right: 'I enjoy it taking kids out to real places. This is real in-service, not just sitting around and talking about things to other teachers.' (Teacher)

  In another school, the pupils and the teacher worked on a real project with an engineer. When the practical work was finished, the engineer taught both the pupils and the teacher how to write an accurate and 'real' technical report. This is something that no ordinary in-service training for teachers can do.
- Pleasure at working with sympathetic and interesting adults who were not teachers, but professionals in another area. 'It's good to have new and different views – we tend to become a bit blinkered sometimes as teachers, well, like organisations – you tend to do things the same way because it worked before. Having someone else who is in a very different work space and environment is refreshing, very refreshing.' (Teacher)
- A boost to morale: by working with schools, engineers begin to discover what a complex and professional job teaching is. They relay this message to the teachers and schools they work with and teachers appreciate being valued by other professionals: 'Most of them are very complimentary about the school. That makes you feel good, because the press about teachers is so bad these days.' (Teacher)
- Support with assessment: in some schools, engineers are beginning to become involved in informal assessment procedures. Teachers are finding this supportive and educational. In one school, where the pupils had to design and build artefacts during a problem-solving activity, the engineers were asked to assess the finished products: 'It was wonderful. They were marking things in ways which we wouldn't normally have done.' (Teacher) In other schools, teachers are asking engineers for contributions to pupils' Records of Achievement and Profiles.

hope that the pressures currently on most science teachers will be slightly relieved so that more time can again be found for developing such highly effective activities.

Science fairs and festivals have been arranged by teachers, parents and local scientists to encourage students to share in the fun of doing science, playing with science, and exploring new fascinating phenomena. Science-based companies' open days can also stir the imagination. Even science quizzes, Brain of Britain type competitions and Trivial Pursuits in science can be stimulating and fun if approached in a suitable way.

Many scientists, looking back over their early careers, speak of the influence of the written word on their choice of career. Some book, a biography of Faraday, Rutherford or Darwin, a well-written book on one of the more stimulating aspects of modern science by Dawkins, Gould, Davies or Crick and Watson, perhaps even some science fiction, has fired the young imagination and left a challenging personal impression on the reader.

Magazines like *New Scientist* or *Scientific American* are not beyond the scope of some of our students. Not every student will be receptive to such literature, but many will be and we need to be aware and sensitive to those.

The establishment of a lively library in the science department, or science section in the main school library, and the positive encouragement to read are an important aspect of an effective science department's work. It is helpful to collect articles from newspapers, magazines or industrial companies, in the science department and, by filing them under different headings, encourage students to refer to them for their own work. Time will, of course, need to be allocated either in class or for homework to reading, followed by the opportunity for students to give a plenary review of what they have been reading about to the rest of the class. Students will only take reading scientific articles and literature seriously if they see that their teachers do so, too. It is surprising how much stimulation can be achieved in a science lesson by reference and discussion of an item of scientific relevance from that day's news. And much can be learnt about science and the way scientists work through second-hand evidence; if we limit the students' science experience only to what they can experience at first hand for themselves we will be doing them a serious disservice.

BAAS, BAYSDAYs, Science and Technology Fairs

Each year, since its foundation in 1831, the British Association for the Advancement of Science has held an Annual Meeting to publicise the latest developments and achievements of science to the widest possible audience. These, though serious, are never solemn occasions and seek to present the whole range of science to a variety of audiences of differing ages. Through talks, discussions, exhibitions, visits and activities for young and old, the BAAS 'aims to improve and encourage scientists and non-scientists alike to understand and promote awareness of the importance of science and technology and the way that it affects all our lives'. It is, indeed, a 'science festival', where young people are welcome to rub shoulders with real scientists, hear the leading scientists of the day discuss their latest thinking together, or become involved with the range of hands-on activities provided for them. It provides a picture of science as a dynamic, exciting way of life, with many important questions still unanswered and challenges still to be met. The challenge of such important problems, which have implications both for their environment and future well-being, can stimulate young people in a way that syllabus-bound school science cannot.

BAYSDAYs are similar to the BAAS Science Festivals but aimed specifically at young people, and are held in different centres around the UK. The 1992 BAYSDAY in London attracted 6,000 young scientists to this carnival of science spread among the Natural History Museum, the Science Museum and Imperial College. Talks and presentations by leading scientists focused on the frontiers of science, and on careers in science and engineering. A programme featuring 'Earth and Space' was specifically aimed at the 8–13 age group. There were hands-on activities in the 'Technology at Work' exhibition and egg races, talks and demonstrations. The BAYSDAY Challenge was to build a rocket launcher to hit a target laid out on the floor of the Science Museum. The national finals of the BAYS Masterminds quiz competition attracted a large and excited audience, with a neck-and-neck competition to decide which school would win the Mastermind trophy and £1,000 prize (Box 5.4). Another major attraction was the display of the excellent student projects for the Lucas Prizes, which provide opportunities to travel to international science fairs in Europe and the USA. It was a festival and celebration of high-level science, produced by both students and practising scientists. Nobody visiting that BAYSDAY would fail to appreciate science as an exciting, stimulating and personally worthwhile activity. Boxes 5.3 and 5.4 give something more of the flavour of BAYSDAYs.

Box 5.3 Raindrops keep falling on my head . . .

Each year, as part of our BAYS Masterminds competition, finalists are given 10 days to carry out an extended investigation set by our in-house philosopher, Thales.

Old hands looked forward with great anticipation to finding out what the 1993 Thales Problem would be! Teams approached the exciting challenge with ingenuity and persistence.

THE THALES PROBLEM
It is obvious that raindrops are different sizes in differing weather conditions. Devise and test a method for measuring the size of raindrops.

Did you know that raindrops are not handsomely tapered but often resemble a small hamburger bun? King's School, Worcester, searched scientific literature for that gem of information. They also discovered from the Meteorological Office that raindrop size varies with temperature, atmospheric pressure and humidity. Drops in a light shower are 1–2 mm in diameter and in a thunderstorm 4–5 mm.

With high speed photography, they confirmed the shape of raindrops by freezing the motion of larger drops . . . They measured the speed of drops and also calculated the size.

Just listen to the rain! Calday Grange Grammar School from the Wirral did just that, counting drops electronically using the sound made as each drop hit a metal collecting bowl placed on a loudspeaker connected to an oscilloscope. Several other schools tried this method using different apparatus. Central Newcastle High School used a bowl of water above a microphone to amplify the sound of the raindrop.

A kitchen recipe for success was devised by the eventual winners, Wallingford School. They put a bowl of flour in the 'rain' and then baked it (they didn't say at what temperature!). The resulting small hard balls of baked flour, each representing a raindrop, were weighed to estimate the size of each raindrop.

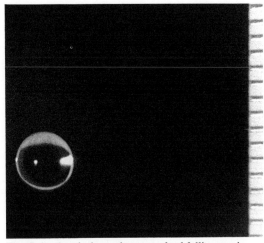

Fig. 5.6 A raindrop photographed falling against a scale marked in millimetres

It's freezing! Wallingford School also put a dish of liquid nitrogen in the 'rain' for a few moments. The nitrogen evaporates leaving frozen raindrops. But you need to be quick, or the drops melt before you weigh them.

Why not try Ashville College, Harrogate's unusual method? You'll need dried cobalt chloride paper, an electronic balance, graph paper on acetate and access to a photocopier.

1 Calibrate the paper by dropping a known mass of water on to it and measuring the area it spreads out to. This will give the mass of water per cm^2 on wetted cobalt chloride paper.
2 Expose the paper to rain – but make sure it's not flooded and that the drops don't merge.
3 Photocopy the paper immediately before it dries out.
4 Find the area of each drop of rain using the acetate graph paper.
5 Convert this to mass using the information from step 1.
6 Convert this to volume by dividing by the density of rainwater (1 g/cm^3)

Box 5.4   BAYS Masterminds Science Quiz

This spring the British Association of Young Scientists (BAYS) held its annual Masterminds Science Quiz, generously sponsored by British Gas. This is organised on a regional basis, with the 16 local winners going to London for the final rounds.

Our school had participated last year and came second in the South Wales heat. The age rules (effectively one pupil from years 9, 11 and 13) meant that we had an entirely new team, so we did not know what to expect. The organisation turned out to be very professional, a media personality as quiz master, buzzers and bells for the speed rounds, some questions based on video tapes and generally a slick presentation. We managed to win the two Cardiff rounds – it all seemed rather unreal – but we had the solid evidence of a ghettoblaster each as solid proof of our success.

The finals were preceded by a very different task which counted as one round in the national semi-final. 'Thales' challenged us to investigate how the bubble size affected the speed at which a bubble rose through a liquid. This was unlike most school experiments as it was open-ended and we were not shackled by school bells. At school the team discussed methods of solving the problem, then I spent the weekend at home doing experiments with my father as an assistant.

We were very excited, rather than nervous, at the prospect of the finals at the Science Museum in South Kensington. Once again we were impressed by the presentation, crisp questioning, clear instructions and keen competition. Our teacher wisely kept quiet his reactions to the tough draw we received, which included the winners from the previous two years. After seven rounds we were just one question ahead of them and they risked an early answer on the last question to catch us up. The question began 'Which carbon allotrope . . .', they guessed diamond, we then confidently said graphite (what else could it be?) but we were both wrong, it was buckminsterfullerene!

Lunch was a quick picnic before the four final teams, fortuitously one from each of the four home counties, were set their second practical task. This was to construct in 45 minutes a rocket, powered by elastic bands, which would land in a metre diameter circle 15 metres away. This project required quick thinking, nimble fingers and teamwork. Although our missile did not land in the circle it was better than some which scarcely left the launch-pad. This challenge counted as one round to the fast and furious final that then followed.

Sir David Attenborough, the questionmaster in the final, had to deal with 12 agile, bright and competitive minds who sought to answer questions across a broad spectrum of scientific disciplines. Topics went beyond the traditional biology, chemistry and physics to include astronomy, geology, safety etc. No team was able to establish a clear lead, indeed when the last question began three of the four teams could still win. 'Where is the intercostal . . .' I buzzed instinctively remembering it as a question that I had wrongly answered in my Mock, but now was certain. 'It is between the ribs.'

We had each won two more items of electronic gadgetry. The school Science Department received £1,000 which was used to purchase an upmarket camcorder. We had learnt some practical investigational science and the need for teamwork, and had enjoyed the experience of participating in a great event.

CHAPTER 6

The way forward

Much of this book has, unashamedly, been a celebration of student research projects and the way they, along with the enthusiasm and expertise of science teachers, have done so much to make science teaching effective for so many students. Student research projects fulfil so many goals, in developing knowledge and understanding, in acquiring scientific skills and experiences and in fostering positive attitudes towards science and the student's own self-image, that it is not surprising that they are seen as central to the whole enterprise of teaching science in schools. Though the overall structured science curriculum that all students experience is important, I would still contend that it is the student research projects (which challenge student's abilities), the stimulus activities in science (which stimulate students' imagination) and teachers' individual relationships with their students (which encourage, support and guide them) that are of primary importance.

Why such projects and scientific investigations are so effective centres on the way they stimulate and develop students' affective attributes, their interest, motivation and commitment, as well as the cognitive aspects. Once students have been 'switched on' they will then be able to acquire the necessary knowledge and skills when required. Without a personal desire to find out more or to follow science further, no amount of knowledge in the head will produce useful outcomes. In the words of Bruner (1966):

> it is only in the trivial sense that one gives a course to 'get something across', merely to impart infor-

mation. There are better means to this end than teaching. Unless the learner also masters himself [*sic*], disciplines his taste, deepened his view of the world, the 'something' that is got across is hardly worth the effort of transmission.

Projects, where students can develop their own motivation and drive, will produce that personal self-confidence that is necessary for autonomous learning so essential in life after school. The project provides the challenge which induces personal commitment to the problem. Working through the investigation, at the level according to students' own potential, will present them with a sense of personal achievement. This personal achievement will build up the self-confidence which will encourage students to tackle even more challenging tasks. The cycle is continuously reinforcing, an iterative process which makes for effective science teaching at the level of each individual student.

It is well known that success in school work and commitment to a particular career are strongly influenced by home background. Our research confirmed this by showing the high proportion of students continuing with science who had parents who had themselves been trained or were working in science. It also reaffirmed the 'fiddle factor': youngsters growing up in homes where they are encouraged to 'fiddle' with mechanical devices, help mend cars, encouraged to share parents' scientific hobbies, and so on, will develop their mechanical skills and aptitudes and are more likely to go into scientific and technological careers. But there is another factor that determines that homes should be such an important and positive factor in children's development, and that is the positive encouragement, the high expectation and the confident support which the parents give their children, all of which builds up children's self-confidence and self-image. We referred earlier to MacKinnon's (1962) important study of successful creative people, in which he analysed how home background was so important for, in this case, creative architects. He concluded:

> What appears most often to have characterized the parents of these future creative architects was an extraordinary respect for the child and confidence in his ability to do what was appropriate . . . [this] appears to have contributed immensely to the latter's sense of personal autonomy which was developed to such a marked degree.

It is this attitude of high expectation which can be transferred to the school context. Indeed, it is exactly the attitude which teachers transfer to their students when they work with them on projects and investigations. Teachers' by showing the expectation that their students will succeed in their projects, produce a self-fulfilling prophecy concerning the students' confidence and success. So through such projects, we can replicate in school those factors which in a supportive home prove so effective. Teachers can thus provide the secure structure in which a challenge attracts commitment, leading to achievement and self-confidence, which in turn leads to renewed ability and desire to face new challenges (Figure 6.1).

Fig. 6.1   The affective cycle for autonomous learning

Of course, such projects cannot be tackled in an intellectual vacuum; the science curriculum needs to be so structured as to provide the grounding and the stimulation for personal investigations. I have stressed in this book the need to develop commitment and confidence, and the need to utilise the whole of the student's experience, whether tacit or explicit, because it is so easy to overemphasise the other factors. Many describe their science curriculum in terms only of the knowledge that the students must explicitly acquire and the scientific skills that they should demonstrate. I wish to stress also the tacit knowledge and the craft skills that practising scientists use in their daily work, and the motivation, feelings and self-confidence that drives them to it.

> Scientific work is necessarily a craft activity, depending on personal knowledge of particular things and a suitable judgement of their properties. (Ravetz, 1971)

> Believing is where learning starts. We know first, act on such knowledge and then get to know more . . . the activity of getting to know is compounded of feelings as well as intellectual curiosity, of hunches as well as facts. (Hodgkin, 1985)

I wish to stress this holistic nature of the scientific enterprise, in contrast to the reductionist approach in which learning science becomes simply the acquisition of a number of scientific facts, principles, theories and skills – each of which when set into its natural context of scientific activity is important but which, when separated and isolated from it, becomes sterile. Effective science teaching needs to integrate all of these attributes, and one of the most genuinely scientific ways of doing this is through student research projects (see Figure 6.2).

However, having stressed the centrality of such individually stimulating science research projects it is necessary to consider the implications for the wider science curriculum throughout the school. At the beginning of this book various issues were raised and unresolved tensions discussed in relation to the science curriculum as a whole. It is now time, in the light of the subsequent discussion,

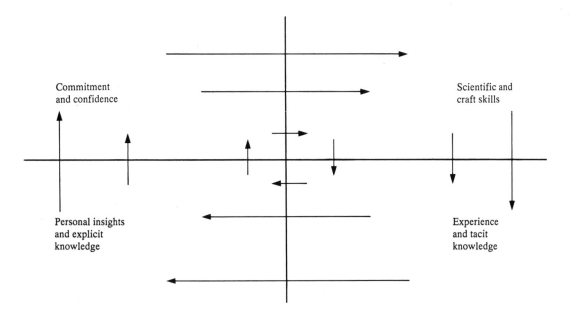

Fig. 6.2 Components of scientific activity

to try and bring these together in the hope that they will illuminate the way forward.

In England and Wales recently, the government has been producing a national curriculum which spells out in great detail what needs to be taught and what needs to be assessed. Such national curricula are already familiar to many other countries, though the detailed nature of the model employed in England and Wales, the degree of prescription and the assessment framework imposed upon the whole curriculum, do come as an (often incredulous) surprise to many from other countries used to a more benevolent and broad-brush approach. However, the existence of such a centrally imposed curriculum does not obviate the need for teachers to think through for themselves the principles behind their own science teaching. On the contrary, teachers need to be all the more aware of their vital role in the 'delivery' of such a curriculum, for it is they who mediate the 'given' curriculum to the students, it is they who decide how it shall be taught, what teaching and organisation styles are most effective, and how the students will experience and enjoy science.

Though science teachers in England and Wales are currently much overburdened by their National Curriculum and, for political reasons, dominated by the demands of its assessment system, one must hope and anticipate that this unreasonable pressure will be reduced in time, allowing more space for teacher initiative. In the government's drive for school and teacher accountability, the assessment system insists that all students are assessed against a very sophisticated, Byzantine structure which has reduced the syllabus to a mass of supposedly criterion-referenced assessment objectives. Hence teachers are obliged to provide evidence for students' attainment against these attainment targets and, consequently, teach to the tests. One effect of the introduction of the National Curriculum has been the vast increase in time that teachers have been obliged to spend in assessing and administering the result of those assessments of students' work, with the consequent decrease in time spent teaching science to them, whether in the normal lessons or outside the curriculum. This was particularly evident in the research described in Chapter 2, which found distinct evidence of a reduction in the quality of science provision, partly through imposed demands and partly as a result of 'innovation fatigue', of reduced teacher time and energy. Already, however, there is considerable variation between schools in the way they respond to the strictures of the National Curriculum in science. For some it is a strait-jacket restricting teachers' creativity. For others it is providing a useful framework around which they can develop their own style of teaching science. We must anticipate that as the system settles down and is simplified, teachers will increasingly use their own professional judgement to interpret the curriculum and teach it in a way that genuinely enhances science education for their students.

In the remainder of this book we will consider the implications for ways forward for effective science teaching under two interrelated and overlapping headings: types of curriculum; and types of organisation and teaching. In a sentence, we will be looking for ways to produce a relevant, stimulating, accessible science curriculum which is appropriate for the vast majority of, if not all, students, which will allow for treatment at different depths according to the students' potential and which has sufficient space to allow different students and teachers to develop their own research projects. In Chapter 1 we discussed various tensions in the science curriculum with various alternatives, vocational or educational aims, rigour or accessibility, content or way of working, doing or thinking, qualitative or quantitative, mathematical or descriptive, teacher directed or student centred, extrinsic or intrinsic motivation. I would suggest that, for most of these issues, the way forward lies not in choosing between one or the other alternatives but in both. If we want to provide a curriculum which meets the wide range of interests and proclivities among, and within, our students we will need to provide them with a varied menu, and the space and flexibility for all to take responsibility for more of their own learning.

Curriculum issues

If we take, initially, the narrow meaning of the word 'curriculum', 'that which describes the syllabus and the content of the course to be taught', there are certain principles which need to be applied. The curriculum needs to be planned and structured in such a way that coherence, continuity and progression can be established. Such a structure needs to be used nationally to ensure continuity for students moving between schools; hence the widely accepted belief in the need for a national curriculum. It is sensible that students should not repeat topics at the whim of the teacher, but rather that if a topic is to be revisited on educational grounds to develop it more fully then a shared awareness of what has been done before will enable constructive reinforcement and efficient development of the topic. Too often in the past teachers have disregarded the science that students have studied in a previous school (especially secondary school teachers of any science that students may have done in their primary schools) on the grounds that there has been no consistent pattern followed in different primary schools and thus no uniform pattern to build upon. It is sensible that progression is sought in the development of a topic and that a student aged 15, say, should have progressed beyond the level of understanding, achievement and practice in a topic that would have been achieved at the age of 11.

Continuity and progression, therefore, are self-evident goals that should be sought from any science curriculum. Having said which, however, it should be recognised that whereas continuity and progression can look straightforward on paper, in reality the way children develop and learn science makes such concepts very variable and imprecise. It is not easy to predict that some topics are harder than others, for all students, and thus should come later. It is not logical to say that two different topics are equally difficult and thus should be at the same level. It is not meaningful to define a precise development of understanding of a given concept which is strong enough to withstand a number of different tests – 'understanding' is a word which covers a very wide range of interpretations. It is not sensible to define in tightly prescribed ways progression in the quality of a student research project or a scientific investigation which may develop unpredictably. All of which is not to say that progression is not an important concept, merely that, like all important concepts (like love, beauty and friendship), it cannot be measured precisely or described in absolute, universally meaningful ways. Treated lightly and with due scepticism, continuity and progression can be important concepts supported by a well-structured national curriculum. Taken too seriously, too analytically and too ingenuously, they can impose an artificial construct upon science teaching which inhibits rather than encourages growth. Certainly the current (1993) preoccupation with measuring levels of attainment in all pupils in all topics in all years, imposed by the National Curriculum for England and Wales, has been as counter-productive as pulling up plants to see how they are growing usually is. In the words of a wise sceptic: 'you don't make a pig heavier by weighing it!'. So we must take note of continuity and progression but not take them too seriously or believe that they actually relate with any precision to what is going on in a student's mind at a particular time.

The curriculum should be representative of the major aspects of science, relevant to students' lives and abilities, and comprehensive enough to give students an appreciation of those areas which have historically and in modern life been important in the development of society. It should help students appreciate and enjoy the world in which they live, and through understanding feel both a sense of enpowerment in the world and due humility concerning their places in it. It should also be sparse enough to allow space for students and teachers to develop their own interests, and to consider issues which have particular local or contemporary interest. It would be impossible for all desirable science to be covered in the time available for science even if all that time were to be prescribed. Any selection of topics for inclusion must be, to some extent, arbitrary. Far better, therefore, to be consciously selective, to make that selection representative and significant, but to leave enough space for local concerns and student projects. It is far better to concentrate on the quality rather than the quantity of learning, and to ensure that that learning is genuinely constructive and enjoyable and develops successful experiences and attitudes which will carry students further when left to their own devices.

As to the specific content of a syllabus, research indicates that a wide variety is necessary as different students will find stimulation in different topics. Overall, however, the benefits of fitting science into its human and social context, and showing its relevance to real life, cannot be overemphasised. This not only motivates students more strongly – especially girls who can see little point in playing with apparently pointless toys in experiments or solving some abstract calculation – but also shows science as it really is, a social activity motivated by human purposes.

The perceived difficulty of certain aspects of school science has a certain ambivalence with students. Our research showed that some students were deterred from continuing with science because they found it too difficult, and thus unrewarding. It also showed that some of the brighter students were deterred from continuing with science because they found it too easy, too trivial, and thus unrewarding! Though the need for accessibility must be addressed, so that the ideas and topics are made accessible to all the students studying it, we must not so reduce the intellectual demands of the subject so that it becomes trivial and non-scientific. At each stage and for each student the content must be both accessible and stimulating, manageable and rigorous, qualitative and quantitative as appropriate. We must not so dilute the content of pre-16 science courses that they neither stimulate nor provide a good insight into and preparation for the more demanding science courses beyond 16. There are dangers that in the search for accessibility for all, some of the more demanding topics, particularly some of the more mathematical applications of the physical sciences, are omitted. But many future scientists have, in fact, been attracted both by the content and the mathematical approach to problems which they found intellectually satisfying and not difficult. We must ensure, both by the content of the curriculum and the way we organise our teaching, that the potential future scientist, as well as the future citizen, acquires success, stimulation and intellectual challenge from the course, too.

The question concerning whether school science should be pure or applied, academically theoretical or set in the relevant context of the student's world, does – in the UK at least – have a rather dated ring about it. Whereas the school science courses of the 1950s and the first serious curriculum reviews through the Nuffield courses of the 1960s concentrated on the pure science, on acquiring an understanding of the underlying principles of science, possibly with a few applications tacked on to the end, the courses developed during the 1980s and 1990s have taught scientific principles through topics, themes and applications relevant to students' everyday lives and the society in which they are growing up. Statements of policy from the HMI, ASE and the Royal Society have very much encouraged this, as have the national criteria for the GCSE examinations. The course developers, the publishers and the textbook writers have followed this lead and produced a range of splendid science books showing science as permeating all of the modern world. This has had a strong motivating effect on students as the texts have become much more student-friendly.

The only danger with this approach is that it can present science as an eclectic activity, concentrating on a variety of scientific facts and their application in various contexts. It can underemphasise the central, fundamental, unifying principles of science which, with their simple elegance, can be extremely powerful in a wide range of situations and highly appealing to many intellectually maturing students. Again the answer to the question 'is it either/or?' (either pure or applied sciences) must be 'both', with the emphasis shifted according to the student context. The principles,

laws, concepts and discoveries of science are an important part of our culture, as well as useful tools for contemporary problems, and as such should be appreciated and enjoyed by all future citizens. The applications of science in our society, their uses and limitations, and their effect on our society are also important and stimulating to all students and future citizens. The possibility that future scientists study an academic, decontextualised science course while the non-scientists study a 'science in society' course is highly dangerous, for future scientists and engineers need to be even more, not less, aware than non-scientists of the implications of science in society. Those who are to use science professionally need, above all, to be aware of its social consequences. Unfortunately, in many countries, the science that is studied by the science specialists, in school beyond the age of 16 and in universities and colleges of higher education, is highly academic and unrelated to the lives of the students or the contemporary world in which they will work (Tobias, 1990). Whereas many changes have come to science courses taught to those aged up to 16, much still needs to be done to make post-16 science courses relevant and attractive to students.

The nature of science teaching in its broader social context, with particular reference to the Science Technology and Society movement and to the importance of a wide cultural perspective, are discussed in two other books in this series by Joan Solomon (1993), *Teaching Science, Technology and Society*, and Michael Reiss (1993), *Science Teaching for a Pluralist Society*. It is still possible to teach science as if it were a collection of objective, dehumanised facts unrelated to the real world, and sadly some students still perceive that from their own experience of school science teaching. But there really is no excuse now for not teaching science as a human activity, influencing and influenced by the society in which it is developed, and an important and exciting cultural activity incorporating values as well as facts, personal judgement as well as tidy solutions.

This brings us back to the question whether

science teaching in schools should be focused on educational or vocational goals, and whether the same course can adequately meet the needs of both. In reality, of course, the question is not so much whether as when. Few would suggest that the curriculum for children aged 7 should be determined by vocational considerations, but for most 18-year-olds vocational training of some kind is highly relevant. Undoubtedly, society has a right to expect of its schools that the students it teaches leave them 'fitted' to work in society as well as to enjoy a fulfilled life in society.

However, the question remains concerning what it means to be 'fitted' to work? What do employers really want for their employment? And what is best acquired at school and what in the workplace? Different countries have different organisational solutions to these questions. Some, like Japan, have a highly academic school curriculum and leave the vocational training to the employers. Some, like Germany, have separate schools offering either an academic or vocation curriculum, separating students into the different types of schooling at the age of about 12 depending on their aptitudes and aspirations. Some, like Italy, offer different vocational tracks for different jobs within the same type of schooling. Others, like the old eastern European countries, offered vocational training for all within the common curriculum. In England and Wales we have experimented with the 'different but equal' model for separating students into different types of schooling at the age of 11, grammar schools for the academic and technical schools for the more 'practically orientated' (and secondary modern schools for the rest!), but the peculiarly English preference for knowing rather than doing, for words rather than action, has meant that the status given to any vocational training has been less than to the academic. The general lack of enthusiasm given to the government's recent idea of city technology colleges shows that that bias still remains. David Layton (1993), in his book in this series called *Technology's Challenge to Science Education*, both gives the flavour of the problems of relating science to technology and shows how science and technology can forge a genuinely constructive relationship.

In reality, the main attributes that employers want of any future employee relate not to specific skills, which can be acquired where and when they are needed, but to basic attitudes to work. They will expect a basic level of literacy and numeracy, as well as the broader problem-solving, communication and interpersonal skills. Above all, they will require a positive attitude to work and the self-confidence for autonomous and flexible learning. All of these are developed, as we have seen, through student research projects and the approach to science teaching that we have encouraged here, for all students on educational as well as vocational grounds. So a common science course, up to the age of 16, will go a long way to meeting both the educational and the vocational goals, the needs of the individual student and the needs of society, if it concentrates on stimulating, relevant, accessible science which gives space and opportunity for diversity around students' abilities and aptitudes and encourages freedom to develop their autonomous problem-solving skills through projects.

There will come a time when more specific vocational training will be required, though this can probably best be left until 16. The development of a new structure for the qualifications and training for vocational skills, the General National Vocational Qualification (GNVQ) looks very promising. By focusing on specific student 'competences', which they need to develop and then to demonstrate, and by emphasising the development of student initiative and self-confidence through coursework and extended workplace-based projects, the students are already demonstrating strong motivation and achievement far higher than would have been expected on more prescribed, academic, teacher-directed A-level courses. The emphasis on competences to be achieved rather than standards against which to be measured, as used in the National Curriculum for England and Wales, is proving a much more motivating form of assessment. Though the science structure is still in the developmental stage, there are indications that it will prove a very good basis for effective science teaching for many, previously unsatisfied, 16–18 year-old students.

Organisation and teaching issues

Much of this book has been centred on one type of teaching, the use of student research projects, and the effectiveness of this type of teaching does not need stressing further here. It has stressed also the importance of the role of science teachers, both in the expert and enthusiastic way they teach their science in the classroom and laboratory and in their personal relationships with students, encouraging, supporting and humanising their teaching. But there are other principles about the types of science teaching that are effective and these we will consider here.

Perhaps the first principle relates to the degree in which students are active or passive in their own learning. The principle of active learning is well established, not only among the constructivist psychologists who remind us that students need to construct meaning for themselves from their previous preconceptions and their new experiences. For useful cognitive understanding to be established in students' minds they must have done something with that new information for themselves. In the words of Osborne and Wittrock (1983):

> The brain is not a passive consumer of information. Instead it actively constructs its own interpretations of information, and draws inferences from them . . . to learn with understanding a learner must actively construct meaning. The successful learning of scientists' ideas is as much a restructuring of the way the learners think about the world as it is about the accretion of new ideas to existing ways of thinking.

But, as we have argued earlier, effective science teaching is not only about acquiring cognitive understanding of science, certainly not only about acquiring explicit as distinct from tacit knowledge of the world, it is also about developing attitudes, commitment and enthusiasm for science. It is clear from the words of the students in the research cited in this book and from the accounts of their projects, that the active involvement in scientific activity – both in the establishment of the problem and in taking responsibility for the actual planning and execution of the work – had an important, positive effect on their attitudes, their motivation and achievement. And the key factor is that of student ownership. Once students take responsibility for their own work, it then becomes their problem to be tackled rather than the teacher's problem to be done. The difference in attitude and outcome between a class of students working on their own projects and one which is passively following the teacher's instructions to do the teacher's tasks is considerable, as indicated best by what happens at the end of the lesson. Often, in the latter situation the students chat among themselves when they have finished, while in the former the teacher will have difficulty in clearing the classroom when the lesson is scheduled to end.

The third of the domains that Bloom described, alongside the cognitive and the affective, was the psychomotor domain. It is self-evident that these psychomotor skills can only be developed by using scientific equipment actively involved in scientific activity, and equally self-evident that these are important in any scientific education. To know about science, even to like science, is not sufficient in itself; a student should also be able to do science. A few years ago, in the 1980s, a reductionist approach to the assessment and development of practical skills was propagated in some quarters, when doing science became reduced to performing a series of isolated tasks with scientific apparatus such as reading a thermometer or assembling some glassware. Fortunately, the holistic nature of scientific activity has become more widely recognised as different from and superior to the sum of the parts (Woolnough, 1989). The National Curriculum in England and Wales now encourages and assesses practical work as complete investigations.

Table 6.1 Types of practical work

Exercises	To develop practical skills
Experiences	To gain experience of a phenomena
Demonstrations	To develop a scientific argument or cause an impression
Investigations	Hypothesis-testing: to reinforce theoretical understanding Problem-solving: to learn the ways of working as a problem-solving scientist

Of course, practical work in science does not all have to be investigations. It can also include exercises, which develop specific practical skills; experiences, which introduce students to particular practical phenomenon; demonstrations, which allow the teacher to develop a scientific argument or to create a dramatic impression. And there will be scientific investigations, either of the hypothesis-testing type as currently in the National Curriculum or of the problem-solving type (see Table 6.1). Most student research projects are of the problem-solving investigations type, though they will use and reinforce students' theoretical knowledge. Fieldwork is likely to include aspects of all of these. Each of these types of practical has its place in science teaching, each will be effective if its aim is appropriately targeted and students are active, mentally as well as physically, throughout. Fuller discussions of the role and reality of practical work in school science and of scientific investigations will be found in Gott and Duggan (1994).

One of the most fascinating factors relating to the way students learn relates to the optimum balance between challenge and security. Too much challenge and the student will withdraw, too much security and the student will never grow. One of the interesting findings from the FAS-SIPES research was how much the majority of students appreciated the security they were given by well-structured, teacher-directed science lessons. The majority were not crying out for more freedom or more opportunity to take more re-sponsibility, they welcomed the security of tightly structured lessons, even at times being spoon-fed! Perhaps that is a reflection of the underlying insecurity that many students have in the normal teacher–student relationships of schooling. On the other hand, when given the opportunity for challenge in their work, under supportive conditions, they welcomed and responded well to it. Again, it is not a question of 'either/or' (do students need to be given structure or independence), but 'both'. Students need both security and challenge. Students need enough confidence in their own ability and in the support structures to be able to explore further, to test themselves out and to grow and learn through responding to challenge, as I learnt from my visit to Tokyo. The editorial 'Travel can broaden the mind!' (Box 6.1) was written for *Physics Education* in the context of a physics conference. I believe it to be equally true for all the other sciences too!

I discussed in Chapter 3 Robin Hodgkin's splendid, if unsettling book, *Playing and Exploring: Education through the Discovery of Order* (1985), where he uses the analogy of mountain climbing to illustrate the creative learning style. He speaks of three types of activity in which students can indulge – play, practice and exploring – and suggests that students need to engage in all three to develop creatively. A walker might 'play' in the foothills, enjoy the experience of being among the hills and wish to climb the surrounding hills. He/she would then need to 'practise', to learn in a safe environment the skills of safe rock climbing, rope handling, and so on. But for real competence to be developed the climber must go 'exploring', must tackle frontier areas and push him or herself further than he/she has ever been before. The learning process would involve continual inter-action between these three types of activity as appropriate to the individual student. In terms of science teaching, I would see the 'explorations' as play, the 'exercises' as practice, and the 'problem-solving investigations', the student research projects, as the exploring. Figure 3.1 represents the different phases of such creative learning.

Science teachers have for some time now been

Box 6.1   Brian Woolnough, 'Travel can broaden the mind!', editorial in Physics Education, 22 (1987)

At a recent conference, Professor George Marx from Hungary challenged us to think radically about the nature of an appropriate education for the young people currently in our schools who would have to face an unknown, and unknowable, future. Many of us had received our education in a relatively stable society and most of our teaching had been in an expanding environment, so that we were only recently becoming aware of the 'limits of growth'. The current generation of students will need to manage a declining, degrading environment. 'We inherited a rich planet, we leave them an overpopulated, polluted planet, emptied of fossil fuels and we don't know what to tell them. . . .' Should we pass on to them the traditional physics courses of the past, should we present them with the latest technology and pretend that the answers lie there, or should we concentrate on an education developing the more personal, simple, creative skills and the ability for self-confident decision making? It was clear which of these paths he viewed most optimistically.

This, and my recent experiences, set me thinking about the most effective ways of learning, of learning how to learn, and how to feel confident in an ability to learn to cope in an unfamiliar, alien world and consequently about the most appropriate model for pupil learning. How far is learning achieved through being told, through being led or through finding out independently? My journey from Tokyo airport to the conference hotel, a journey necessitating travel in a bus, a train and then a taxi through a land whose notices I could not read and whose language I could not understand, was simply achieved, as I was taken all the way by a most helpful and hospitable Japanese colleague. However, I quickly realised that I was still not competent to find my own way around this strange city, as I had arrived at my destination without having learnt. It was not until I went out of the hotel on my own, armed with a subway map and a few yen, and started exploring *for myself* that I really began learning how to cope with and feel at home in an unfamiliar land. I travelled slowly and tentatively at first, I got lost and took some inefficient turnings, but these were all a necessary part of the learning process. I suppose there was a slight chance of

getting badly lost and, certainly, of not finding some of the interesting places that only a local would know – though even here a small guide book could help. But the confidence I quickly gained could only have come through being left alone, with a few basic guidelines, the challenges to be met and the personal motivation to explore. It would have been easier, and safer, if I had been led everywhere by a guide. I would have arrived at all the required destinations at the right time, but I would have learnt very little and certainly not developed the confidence and skills required to explore different environments.

The parallels with learning in physics are obvious. Learning, if it is to become meaningful and useful in a wider context, must be individualised and will necessitate risk, exploration and personal commitment. The more we structure our students' learning for them and guide them to ensure that they arrive safely at the right destination, *our* destination, the more they will become dependent on others. The more we can throw the responsibility over to them, to solve their own problems in their own way, the more they will develop their own skills and self-confidence. For practical work this highlights the value of students tackling their own investigations rather than following tightly prescribed experiments. For students to learn physics they must *do* physics, and to learn how to tackle problems as a physicist they must work through their own, open ended, investigations. And through the challenge, the risk, the exploration and the personal commitment this will entail, they will build up their skills and self-confidence.

Does such reasoning lead us to a philosophy of totally independent discovery? Not so, for there will be a need for the pupils to have acquired certain basic skills and some knowledge before they can start investigating confidently. I needed a map, and the competence to read a map, I needed to know that there was a subway system and where to buy the tickets and I needed sufficient self-confidence and motivation to reach my destination before I could travel around Tokyo successfully. Our problem is to know how much knowledge and which skills our pupils need and how we can build up their self-confidence and their motivation before letting them loose 'on their own'. I suspect the answer is far less than we currently practise and that most of that knowledge, confidence and skill comes through actually doing investigations and being exposed to appropriate experiences. Learning for an uncertain future is a complex mixture of doing and knowing and, above all, of wanting and being: it is this we must encourage in our students for their explorations in physics.

aware of the different reaction of boys and girls to school science (See Gardner, 1975; Kelly, 1981). In particular, physical science teachers have been aware that many girls dislike and reject the sciences and find much of the practical work trivial and pointless. This again was confirmed in our research. Perhaps the girls are right, there are many practicals in physics and chemistry which are trivial and pointless, and have no relevance to their lives. One striking exception to this is the way girls take with enthusiasm to practicals of the great egg race type or student research projects. They frequently win such competitions and are enthusiastic contributors to other such projects. The reasons are several and point to some important guidelines for all students. Projects will have a clear purpose, often one which relates to areas involving people and the environment, which are important to girls. The girls will be working cooperatively in a team, which girls like, and only in a secondary sense in competition with others. There is time to explore various approaches to the problem before settling on a preferred one – a highly laudable habit which girls often deploy but which puts them at a disadvantage in most standard school practicals which expect a single convergent approach (Murphy, 1991). The projects will lead to a successful outcome, which reinforces self-confidence, in a way that 'the experiment doesn't work' does not.

The project encourages creativity, imagination and divergent thinking, all of which girls respond to well. These personal aspects which are highlighted in such distinctions challenge us to rethink some of the practical work, for all pupils. Do they have a genuine purpose, that students appreciate and share? Do they have any point in the real world? Do they encourage cooperation? Do we allow enough time for students really to think through what they have to do or are students led through them artificially on the route we want them to take? Do they give students a sense of achievement? Do they encourage creativity, imagination and divergent thinking? Perhaps they should.

Another issue which flows from students' reactions to great egg races and such competitions relates to the whole area of the place of competition and assessment in relation to student motivation. When is competition beneficial and when counter-productive? Is extrinsic motivation always detrimental to the possibility of intrinsic motivation? I believe that the positive responses that students give to the competitive element in great egg races, and the way the judges evaluate such endeavours, point important ways forward for assessment and motivation in science.

I believe that there are three principles concerning the effectiveness of competition in school science:

- Competition, as an extrinsic motivator, should only be used as a secondary goal to the intrinsic motivation that comes from the worth and enjoyment of the science activity itself.
- Competition will be a positive stimulus to good work if and only if the student stands a reasonable chance of succeeding.
- If a student has no chance of success in a competition, such competition will be counter-productive

The danger with too much extrinsic motivation, the encouragement of students to work on something because of some extrinsic reward or punishment (such as a mark, an examination grade, a prize, a detention), is that this becomes an end in itself and undermines any possibility of the teacher encouraging the student to take an interest in the activity for its own sake. We all know the student's cry. 'If it's not in the syllabus why are we doing it?'. But somehow, the competition element in great egg races does seem to have a beneficial effect. This is because of the first two principles above. The activity is intrinsically worth-while and enjoyable in its own right, and whether or not the students 'win' the competition they will have succeeded in achieving their goals and found satisfaction in solving their problem. If the competition element gets too great then it does become counter-productive once a group of students realise that they are not going to win.

The other aspect of such competitions, and also the CREST award type schemes, is that the assessment is based on broad, sensible, comprehensible criteria, with final evaluation being made by an expert's professional judgement. This allows not only for judgement against the specific criteria of the competition's goal (for example, 'which bridge supports the greatest weight?'), but also some credit for more general design features such as 'degree project has utilised scientific knowledge' or 'originality of design' or even 'aesthetic appeal of construction'. The aspect of professional judgement is not only recognised as fair but also much more real to adult life. There, such judgements are used continuously in assessing the suitability of an applicant for a job, the best design for a building, the best performance of a piece of music, even the best dress for a party. Important criteria have to be left to personal, professional judgement, they cannot be measured against tight, prescribed, objective, entirely reliable criteria. Professional judgement based on broad criteria is a far more acceptable and valid form of assessment than one based on tight predetermined criteria. Such prescriptive, objective, 'teacher-proof' assessment schemes cannot make any judgement on many of the more important aspects of a student's work, such as flair, creativity, persistence and 'consumer-friendliness'.

In many ways the ability of a student to do good scientific work is akin to a student's ability to write a poem or paint a picture, the quality of which can only be judged by an 'expert' observer. We must find a way forward for assessing students' work such that it will be assessed in sensitive, humane ways relying on teachers' professional judgement against a set of broad criteria. We must continue to stress the need for validity to take precedence over reliability in our assessment. For only then will the quality of students' scientific work improve in significant ways, and teachers' own assessment techniques and teaching of valid science be developed, too. Fortunately schemes such as the assessment of Nuffield A-level physics (Jakeways, 1986) and the use of broad-cued report sheets (Woolnough and Toh, 1990) give important clues to effective forms of assessment.

So far, the issues relating to the way we teach have concentrated on broad principles and have not stressed specific teaching strategies. The question as to what teaching strategies research has shown to be most effective has not yet been addressed. Such an apparently simple question, however, needs to be qualified by 'effective for what?' and 'effective for whom?'. As discussed in Chapter 1, there is a variety of aims for teaching science and different teaching strategies will be more or less effective in meeting each of them. Creating a poster display will increase a student's communication skills. Working through a scientific calculation will reinforce a student's grasp of a scientific principle. Doing a scientific project will teach a student how a scientist works.

Furthermore, students themselves are different and our research has illustrated that they will react differently to a particular teaching strategy. Some students love mathematical calculations and find them stimulating, others hate them and find them a barrier to comprehension. Some students gain confidence and insight by doing practical experiments, others find them unnecessary and the surrounding 'clutter of experimentation' distracting to the underlying principles. Some students find writing up their own experiments helpful in 'writing themselves into understanding', others find too many words a barrier and communicate best through diagrams and cryptic phrases. Some students respond best to the security of tightly structured lessons, others prefer to have more freedom to develop their own ideas and to make their own mistakes!

At quite deep levels students differ in the way they think, work and perceive the world (Head, 1979): some tackle problems in a serialist way, others in a holistic way; some 'see' the world in pictures and models, some in words, some find mathematics sufficient; some give priority to the relationship to people, some prefer dealing purely in ideas and abstract concepts; some are concrete thinkers, some abstract; some are risk-takers, others prefer to play safe; some work best on their own, others sort out their ideas in discussion with others; some enjoy competition, some prefer co-operation; some work best under pressure, others excel when working at their own pace; some have scientifically supportive home backgrounds, others learn only that science that they meet in science lessons; some love fiddling and playing with gadgets, others find such mechanical 'toys' either trivial or frightening; some will find academic work easy, some will find it hard; some utilise the explicit knowledge learnt in school, others have a wealth of tacit knowledge from their hobbies and home background which can be invaluable.

In a class of 30 pupils, there will be a spread of most of these variables! It seems a daunting, if not an impossible task to find teaching strategies that satisfy them all. Clearly a perfect matching exercise is impossible. It is useful to recognise that and settle for what is feasible. The best solution is to provide a variety of teaching strategies throughout the course, so that different students will benefit from different strategies in different lessons, and allow enough freedom and flexibility so that each student can, when motivated, take responsibility and ownership of his/her own learning style. This again drives us back to the importance of the affective domain in effective science teaching. If students are motivated, and if they are given the freedom and opportunity, they will find ways of learning. If they are not, they will not bother.

Keith Postlethwaite's (1993) book in this series, *Differentiated Science Teaching*, gives a fuller discussion of ways of responding to individual differences and to special educational needs. As to the range of teaching strategies, readers should consult the book by Bentley and Watts (1989), *Learning and Teaching in School Science: Practical Alternatives*, where they and fellow teachers describe a wide range of different strategies in practice. Throughout the book they stress the importance of the students being active in their own learning – 'learning to make it your own'. They discuss practicals and projects, talking and writing for learning, problem-solving, encouraging autonomous learning, games and stimulations, using role play, and media- and resource-based learning. In all, they give 28 case studies of these different strategies in action – a veritable mine of good teacher-tried experience.

We have spoken at length about the value of practical work of different types in teaching science. In the UK practical work of some kind has always played a large and central part in science teaching. Indeed, some teachers almost feel guilty if they are not doing practical work in their science lessons! But science is only partly about doing; at its heart science must be about knowing and understanding the concepts and principles of science. This requires hard thinking, discussing, applying and reordering the cognitive structures in the mind. Above all, science is about the use and meaning of words, the meaning given to language and the models and analogies that scientists use to understand and 'make sense' of the world. Such scientific thinking does not come easily, and practical work can, if we are not very careful, be an irrelevant, time-consuming, distraction from the exercise of refining students' thinking. The more direct value of thinking, talking, discussing, writing and applying ideas in science lessons cannot be overestimated. This is not the place to discuss them in full, readers should consult Clive Sutton's (1992) delightful and revolutionary book in this series, *Words, Science and Learning*. Joan

Solomon's (1993) books, *Teaching Science, Technology and Society* also illustrates, through her researches on student learning, how students construct their own knowledge through social intercourse and discussion.

The issue of whether science teaching in schools is best organised through the separate sciences of biology, chemistry and physics, or as a single subject is one which has concerned science educators and head teachers for some time. In practice, solutions have often been influenced as much by administrative convenience and staffing expediency as by educational considerations, but it is worth considering what is preferable and whether research evidence indicates one way as preferable to another.

The first question relates to the different age ranges of the students, and to when the different sciences grow so distinctive that the differences between, say, biology and physics outweigh what they have in common. Clearly, when children are first introduced to science at around the age of 5 or 6, the emphasis is on science as a way of exploring the world, and knowledge is largely factual and accessible to all children and teachers. At more sophisticated levels, when students are specialising in science at the age of 18 or in higher education, the differences between biology and physics are quite distinct. Not only is the content knowledge quite extensive and distinct, but the way a physicist and a biologist approach their subject is different, too. Many students like one approach and dislike the other. At this stage, the distinct culture of the subjects suggests that it is more sensible to teach them as separate subjects. So the question becomes, at what stage between the ages of 5 and 18 it is more effective to move science teaching from the single-science organisation to the separate sciences.

Much of what our research has shown about successful science teaching applies equally well to both forms of organisation. Indeed, all of the earlier discussion in this chapter, about the issues affecting the curriculum content and the ways of teaching science, applies equally to science as a single subject or to the teaching of biology,

chemistry and physics separately. In the Netherlands, for instance, all the Science, Technology and Society (STS) developments have been developed more through the PLON physics curriculum whereas in the UK, STS has been developed through an integrated science approach. Clearly, active learning methods, student research projects, stimulus activities in science, extra-curricular science activities, science clubs and competitions, and school–industry links can all be organised through the separate sciences or within a single science approach. The important role of the science teacher, both in enthusiastically teaching his/her own discipline and in encouraging and supporting students, could equally well be organised through science or the sciences.

It is, however, in the role of the teacher rather than of the subject that the benefits of separation begin to emerge. How far can individual science teachers enthuse and be expert in all areas of science? How much time need hard-pressed science teachers spend in becoming expert in a field of science outside their own graduate specialism and will that time reduce the time they might have spent in, say, organising extra-curricular activities or talking to the students? Clearly the answer to such questions will be different for the teachers in different science departments. But it is the answers to such questions that should determine how science is organised, so that each department can utilise the expertise, the energy and the time of their staff to the best advantage. Our research showed that most science teachers feel happy teaching the sciences as a single science up to the end of Key Stage 3, the age of 14. But it also showed that many teachers felt unhappy about their competence to teach all three sciences expertly, enthusiastically and effectively into Key Stage 4, 14–16, and beyond. Some felt well able to do so, but many teachers believed that they were teaching most effectively when they 'were teaching their own subject' at this stage. Physicists and chemists resented 'having to mug up on all the biological facts', biologists were unhappy coping with some of the more abstract concepts in physics and chemistry, and felt unprepared to discuss their applications fully. They felt that the extra work-load required to tackle the new topics could have been better spent teaching their own subject to the students; for some teachers it was 'the last straw'.

The teaching profession rightly expects its teachers to be expert in the subject they are teaching, for teachers of a subject up to the age of 16 to have graduated in that subject. We expect maths graduates to teach maths and history graduates to teach history. The suggestion that a biology graduate can teach chemistry or a physics graduate teach biology – 'because they are all one science really' – shows an ignorance and disregard for the nature of the sciences that could only be countenanced by non-scientific heads and administrators. If the enthusiasm of science teachers for their subject is so important in successful science teaching, as our research has shown, then it is not effective to organise science teaching in such a way that such enthusiasm and expertise are dissipated.

A counter-argument relates to the time that teachers spend with their students, the time that it takes for teachers and students to get to know each other and work most effectively together. If we split science teaching up into, say, three periods a week and have one period per science taught by three different teachers, as might well be done if we insisted on teaching physics, chemistry and biology as separate subjects in Key Stage 3, teachers would not have sufficient time to get to know their pupils and thus teach them effectively. Where the teaching allocated to science is 20% of the curriculum time in Key Stage 4, say five hours per week, it is slightly more feasible to split the teaching into distinct subjects with teachers teaching to their own expertise.

The two most common ways of organising this are through parallel, concurrent, coordinated science courses with the three separate sciences each having, say, two 50-minute periods per week for two years, or through end-on, consecutive, modular science courses with the different modules taking about a ten-week period, and teachers changing classes at the end of each module. The strength of the former is that teachers can teach their own subject specialism to the same class for two years; the strength of the latter is that pupils spend more time each week on a particular topic and get more short-term reward by assessment at the end of each module. The problem with the former is that there is little teacher–student contact each week; the problem with the latter is that teachers do not have continuity with their classes throughout the two years, and thus waste time at the beginning of each module getting to know the students and the level of their scientific attainment and potential. I suspect that there is more future in the coordinated approach than the modular approach in Key Stage 4, because of the advantages of continuity throughout the course.

Two of the questions we asked the heads of science in their questionnaire in the first phase of the FASSIPES research concerned how far they encouraged their students to take three separate sciences in years 10 and 11 and the extent to which integrated science was taught up to year 11. We found that there was a very strong, positive, correlation between the 'success' of the school and the extent to which they encouraged the three separate sciences, and a strong, negative, correlation between the 'success' of the school and those who taught integrated science. 'Success' was measured solely in terms of the proportion of students going on to study one of the sciences or engineering at university. This, of course, is only one measure of effective science teaching and one which relates to a minority, but not an insignificant minority, of the school population. Caution needs to be given to this finding, too, in that it recorded a correlating, not a causal link. There is evidence that schools with a strong tradition of science teaching have well-qualified staff in each of the three subjects and have able students for whom the separate science courses were originally designed. Some of the schools with weaker traditions in science teaching often have less academically able students and have more difficulty recruiting a well-balanced and well-qualified science staff. Thus for administrative convenience as well as the availability of integrated science courses for the less able student they are less likely to be able to promote the three separate sciences and more likely to encourage integrated science. Our subsequent research unearthed the causal link between the organisational arrangements and the 'success' of the school, through the enthusiasm and expertise of the science teachers being used most efficiently.

Each school needs to sort out its own organisational pattern according to how it can most efficiently deploy the expertise and enthusiasms of the staff that it has at its disposal. It is not the curriculum but the science teachers themselves who are the key to effective science teaching.

Stop press! The afternoon after I had written the section on organisation and teaching issues, I came across the following 'confession' in an educational newsletter (Ford, 1993). I reproduce it without further comment (see Box 6.2).

But is it possible?

I have been conscious as I have been writing that I have been celebrating and encouraging things which may take extra time – student research projects, involvement with extra-curricular activities, and taking time to talk with students and teach with enthusiasm. I am also conscious that teachers in England and Wales are currently under enormous pressure of work to teach, assess and administer the National Curriculum, are being directed by that curriculum towards a different, reductionist, prescribed type of science teaching and are overloaded with 'innovation fatigue'. I am conscious that many science teachers in other countries do not have the tradition of, or facilities for, much practical work in school laboratories. Is what I have been advocating realistically possible? Can it be fitted in with all of the other constraints? I believe the answer to that is that it can, indeed must if effective science teaching is to continue to develop effectively.

My optimism in the possibility of such activity in science is based primarily on the fact that some teachers are currently doing it already. This book is not just an exhortation of what might happen but

---

Box 6.2   A heretic confesses

I am about to make a statement that is reactionary, unfashionable heresy. I am a Biology teacher, I am not a Science teacher. I enjoy teaching Biology; I want to give young people some understanding of the fascinating and intricate ways in which their bodies work and the wonderful variety of living organisms and the complex and subtle interactions between them. I believe that I can teach Biology well. I can teach Chemistry adequately but not well – I find much of it rather dull. I teach Physics badly – to me much of it is either very boring or totally incomprehensible, or both. Given a textbook and a stock of work sheets I can teach Science, but I can't enthuse over the Chemistry and Physics based topics. I haven't got a store of old facts, quirky anecdotes and silly jokes to liven up dull areas. I have an insufficient depth of knowledge to lead an academic child further in the subject, and I cannot produce simple metaphors to help less able children understand. I probably make many errors. I cannot answer children's questions on Physics and Chemistry topics with any confidence; if they ask me questions about magnetism I know little more than what is printed in the textbook in front of them. But if they ask me about diabetes, frogspawn development, Siamese twins, adder bites, cancer or tapeworms (all topics of great interest to children) I can answer them with assurance.

---

a celebration of what is happening. I have tried to include throughout illustrative activities which students and teachers are currently engaged in, despite the pressures and constraints acting on them. If some can do it, my hope is that more will do it. I recognise the unhelpful pressures of the National Curriculum: I trust and believe that by the time this book is published the science curriculum will be simplified in such a way that science teachers will have more time and space to breathe, to teach science and still have enough energy left for extra-curricular activities in science. I recognise that many teachers have limited apparatus and laboratory facilities, but in fact many, perhaps most, of the best students' projects require little sophisticated equipment, especially if based on those most fertile of all areas – the environment and the home. Professor Ernest Rutherford is reputed to have said to his colleagues at the Cavendish Laboratories in Cambridge: 'We haven't the money so we will have to think!'

Since writing this draft, and just before the manuscript finally went to press, Sir Ron Dearing (1993) published his interim report on *The National Curriculum and its Assessment* for the UK government's Department for Education, which has accepted all his recommendations. This is excellent news, as Sir Ron has looked at the National Curriculum and the way that it is being implemented, and has shouted 'halt!'. The structure needs to be simplified, the curriculum needs to be slimmed down, the assessment needs to be more holistic, the teachers need to be trusted more to use their professional judgement, and the administration and central organisation of the implementation need to be more efficient and match the schools' timetables. He has picked up the worries than many of us had had about the detrimental effect of some aspects of the National Curriculum on science teaching, many of which I have expressed in this book. Most important, he has recommended that the work overload which the present structure has been imposing on teachers should be greatly reduced. So, in future, science teachers will not have every moment of their time taken up in 'delivering' and assessing the National Curriculum. They will have more freedom to use their professional judgement, to talk more with their students, to spend more time on quality teaching and less on assessment, and, most important of all, to be able to introduce more student research projects and other stimulus activities in science.

Ultimately, the quality of science teaching depends not on the given curriculum but on the quality, the vision and the energy of science teachers themselves. It is the teachers, not the politicians, who mediate the curriculum and work with the students. As these teachers catch a vision of the potential of students for doing high-quality science work through projects, and as they confidently sort out their own priorities to facilitate these and other extra-curricular activities, they will find, as many of us have done before, that science teaching can be not only effective but also thoroughly satisfying.

Useful addresses

Association for Science Education. College Lane, Hatfield, Herts AL10 9AA

British Association (BA and BAYS). Fortress House, 23 Savile Row, London W1X 1AB

Chemistry Club. Homerton College, Cambridge CB2 2PH

CREST. The Technology Centre, University Research Parks, Guildford, Surrey GU2 5YG

Engineering Education Scheme. Astwick Manor, Cooper Green Lane, Hatfield, Herts AL10 9BD

Institute of Biology. 20 Queensberry Place, London SW7 2DZ

Institute of Physics. 47 Belgrave Square, London SW1X 8QX

Neighbourhood Engineers. The Engineering Council, 10 Maltravers Street, London WC2R 3ER

Royal Society. 6 Carlton Terrace, London SW17 5AG

Royal Society for Chemistry. Burlington House, Piccadilly, London W1V 0BN

Science and Technology Regional Associations. c/o Standing Conference on Schools Science and Technology, 76 Portland Place, London W1N 4AA

The TASTRAC project. TASL, H.H. Wills Physics Laboratory, Bristol BS8 1YR

References

Allen, J. E., Camplin, G. C., Henshaw, D. L., Keitch, P. A. and Perryman, J. (1993) 'A UK national survey of radon in domestic water supplies', *Physics Education*, 28 pp 173–177.

Assessment of Performance Unit (1989) *Science at Age 11 (and 13, and 15)*. London, HMSO.

Ausubel, D. (1968) *Educational Psychology, a Cognitive View*. New York, Holt, Rinehart and Winston.

Barnes, D. (1972) *From Communication to Curriculum*. London, Penguin.

Bentley, D. and Watts, M. (1989) *Learning and Teaching in School Science: Practical Alternatives*. Buckingham, Open University Press.

Black, P. (1990) 'APU Science – the past and the future', *School Science Review*, 72: 13–28.

Breakwell, G. M., Fife-Schaw, C. and Devereaux, J. (1988) 'Parental influence and teenagers' motivation to train for technological jobs', *Journal of Occupational Psychology*, 61.

Bridges, D. (ed.) (1992) *Neighbourhood Engineers Source Book*. London, Engineering Council.

British Association for the Advancement of Science (1982) *Awards for Young Investigators*. London, BAAS.

British Association for the Advancement of Science (1983) *Ideas for Egg Races*. London, BAAS.

British Association for the Advancement of Science (1985) *More Ideas for Egg Races*. London, BAAS.

Bruner, J. (1966) *Towards a Theory of Instruction*. Cambridge, MA, Belknap Press.

Bryce, T. G. K. and Robertson, I. J. (1985) 'What can they do? A review of practical assessment in science', *Studies in Science Education*, 12: 1–24.

Bryce, T. G. K., McCall, J., MacGregor, J., Robertson, I. J. and Weston, R. A. J. (1983) *Techniques for the Assessment of Practical Skills in Foundation Science*. London, Heinemann.

Campbell, B., Lazonby, J., Millar, R. and Smyth, S. (1990–) *Science, the Salters Approach*. London, Heinemann.

Camplin, G. C., Henshaw, D. L., Lock, S. and Simmons, Z. (1988) 'A national survey of background alpha-particle radioactivity', *Physics Education*, 23: 212–17.

Clackson, S. G. and Wright, D. K. (1992) 'An appraisal of practical work in science education', *School Science Review*, 74(266): 39–42.

Clark, A. (1993) 'The Engineering Education Scheme: a school's point of view', *Physics Education*, 28: 271–3.

Claxton, G. (1991) *Educating the Enquiring Mind: the Challenge for Schools*. Hemel Hempstead, Harvester Wheatsheaf.

Coles, M., Gott, R., Price, G. and Thornley, A. (1988–91) *Active Science Books 1–3*. London, Collins Educational.

Davies, K. (n.d.) *In Search of Solutions: Some Ideas for Chemical Egg Races*. London, Royal Society of Chemistry.

Dearing, Sir Ron (1993) *The National Curriculum and Its Assessment, Interim Report*. York, NCC.

DES (1991) *Science in the National Curriculum*. London, HMSO.

Devlin, T. and Williams, H. (1992) 'Hands up those who were happy at school', *New Scientist*, 26 September: 40–2.

Dobson, K. (1992) 'Creating a monster in the lab', *Education Guardian*, 31 March.

Driver, R. and Bell, B. (1985) 'Students thinking and the learning of science: a constructivist view', *School Science Review*, 67: 443–56.

Eggleston, J. (1992) *Teaching Design and Technology*. Buckingham, Open University Press.

Engineering Council (1992) *Neighbourhood Engineers: Practical Support for Schools*. London, Engineering Council.

Engineering Education Continuum (1992) *Engineering Talent for the Future*. London, Royal Academy of Engineering.

Ennever, L. and Harlem, W. (1972) *Science 5–13*. London, Macdonald.

Fensham, P. J. (1985) 'Science for all: a reflective essay', *Journal of Curriculum Studies*, 17: 415–35.

Ford, B. (1993) 'A heretic confesses', *CiSE Newsletter*, 5: 5.

Galloway, J. (1991) 'Working class honours: the not so glittering prizes', *New Scientist*, 28 March.

Gardner, P. L. (1975) 'Sex differences in achievement, attitudes and personality of science students, a review', *Science Education; Research 1974*: 231–58.

Gee, B. and Clackson, S. G. (1992) 'The origin of practical work in English school science curriculum', *School Science Review*, 73(265): 79–83.

Gott, R. and Duggan, S. (1994) *Investigative Work in Science*. Buckingham, Open University Press.

Head, J. (1979) 'Personality and the pursuit of science', *Studies in Science Education*, 6: 23–44.

Head, J. and Ramsden, J. (1990) 'Gender, psychological type and science', *International Journal of Science Education*, 12(1): 115–21.

HMI (1985) *Curriculum 5–16. Discussion paper 2*. London, HMSO.

Hodgkin, R. A. (1985) *Playing and Exploring: Education through the Discovery of Order*. London, Methuen.

Hodson, D. (1990) 'A critical look at practical work in school science', *School Science Review*, 70(256): 33–40.

Hodson, D. (1992) 'Redefining and reorientating practical work in school science', *School Science Review*, 73(264): 65–77.

Hutchinson, B. (1993) 'The Engineering Education Scheme', *Physics Education*, 28: 267–70.

Jakeways, R. (1986) 'Assessment of A level Physics (Nuffield) Investigations', *Physics Education*, 21: 212–14.

Jenkins, E. (1979) *From Armstrong to Nuffield*. London, John Murray.

Joint Council for 16+ National Criteria (1981) *Draft National Criteria for Science*. Manchester, Joint Matriculation Board.

Kalmus, P. I. P. (1985) 'What attracts students towards physics', *Physics Bulletin*, 36: 168–71.

Kelly, A. (ed.) (1981) *The Missing Half: Girls and Science Education*. Manchester, Manchester University Press.

Kelly, A. (1987) 'Special Issue: Gender and Science', *International Journal of Science Education*, 9(3): 259–417.

Layton, D. (1973) *Science for the People*. London, Allen & Unwin.

Layton, D. (1993) *Technology's Challenge to Science Education*. Buckingham, Open University Press.

Lister, J. M. (1994) *Chemical Demonstrations for Schools*. London, Royal Society of Chemistry.

Lock, R. (1988) 'A history of practical work in school science and its assessment 1860–1986', *School Science Review*, 70(250): 115–19.

MacKinnon, D. W. (1962) 'The nature and nurture of creative talent', *American Psychologist*, 17: 484.

Medawar, P. B. (1969) *Induction and Intuition in Scientific Thought*. London, Methuen.

Millar, R. (1991) 'A means to an end: the role of processes in science education' in B. E. Woolnough, (ed.), *Practical Science*. Buckingham, Open University Press: 43–52.

Misslebrook, H. (1971) *Nuffield Secondary Science, Teachers' Guides*. London, Longman.

Murphy, P. (1991) 'Gender differences in pupils' reaction to practical work' in B. E. Woolnough (ed.), *Practical Science*. Buckingham, Open University Press: 112–22.

Nuffield (1967) *Nuffield Junior Science*. London, Collins.

Osborne, R. J. and Wittrock, M. C. (1983) 'Learning science: a generative process', *Science Education*, 4: 489–508.

Polanyi, M. (1958) *Personal Knowledge*. London, Routledge and Kegan Paul.

Pope, M. and Watts, M. (1988) 'Constructivist goggles, implications for process in teaching and learning physics', *European Journal of Physics*, 9: 101–9.

Postlethwaite, K. (1993) *Differentiated Science Teaching*. Buckingham, Open University Press.

Power, C. (1977) 'A critical review of science classroom studies', *Studies in Science Education*, 4: 1–29.

Ravetz, J. (1971) *Scientific Knowledge and its Social Problems*. New York, Oxford University Press.

Reiss, M. (1993) *Science Education for a Pluralist Society*. Buckingham, Open University Press.

Roe, A. (1953) 'A psychological study of eminent

psychologists and anthropologists and a comparison with biological and physical scientists', *Psychology Monographs*, 67(2).

Rogers, E. (ed.) (1977) *Revised Nuffield Physics*. London, Longman.

SATIS (1987–92) *Science and Technology in Society*. Hatfield, Association for Science Education.

Screen, P. (1986) 'The Warwick Process Science Project'. *School Science Review*, 68(242): 12–16.

SEAC (1992) *School Assessment Folder (part 2). Science 1992 National Pilot*. London, SEAC.

Shayer, M. and Adey, P. (1981) *Towards a Science of Science Teaching*. London, Heinemann Educational Books.

Sjoberg, S. and Imsen, G. (1988) 'Gender and science education' in P. Fensham (ed.), *Developments and Dilemmas in Science Education*. London, Falmer Press.

Smithers, A. and Robinson, P. (1992) *Technology in the National Curriculum*. London, Engineering Council.

Solomon, J. (1993) *Teaching Science, Technology and Society*. Buckingham, Open University Press.

Sutton, C. (1991) *Communicating in the Classroom*. London, Hodder and Stoughton.

Sutton, C. (1992) *Words, Science and Learning*. Buckingham, Open University Press.

Swain, J. R. L. (1989) 'The development of a framework for the assessment of process skills in a Graded Assessment in Science Project', *International Journal of Science Education*, 11(3): 251–9.

Taylor, C. (1988) *The Art and Science of Lecture Demonstration*. Bristol, Adam Hilger.

Tobias, S. (1990) *They're Not Dumb, They're Different*. Tucson, AZ, Research Corporation.

Tytler, R. (1992) 'Independent research projects in school science: case studies of autonomous behaviour', *International Journal of Science Education*, 14(4): 393–412.

Tytler, R. and Swatton, P. (1992) 'A critique of Attainment Target 1 based on case studies of students' investigations', *School Science Review*, 74(28): 21–35.

West, A. and Chandaman, R. (1993) 'The real gold standard?', *Physics Education*, 28: 274–83.

Wellington, J. (1981) 'What's supposed to happen Sir?' *School Science Review*, 63: 167–73.

Wiener, M. J. (1981) *English Culture and the Decline of the Industrial Spirit 1850–1980*. Cambridge, Cambridge University Press.

Woolnough, B. E. (1972) 'School–research laboratory liaison', *Physics Education* 7(7): 401–6.

Woolnough, B. E. (1988) *Physics Teaching in Schools, 1960–85*. London, Falmer Press.

Woolnough, B. E. (1989) 'Towards a holistic view of processes in science education' in J. J. Wellington (ed.), *Skills and Processes in Science Education*. London, Routledge.

Woolnough, B. E. (1990a) 'Changes in physics teaching in England since 1960: the people, policies and power in curriculum administration' in H. Haft and S. Hopman (eds), *Case Studies in Curriculum Administration History*. London, Falmer Press.

Woolnough, B. E. (ed.) (1990b) *Making Choices: An Enquiry into the Attitudes of Sixth-Formers towards Choice of Science and Technology Courses in Higher Education*. Oxford, Oxford University Department of Educational Studies.

Woolnough, B. E. (1991a) *The Making of Engineers and Scientists*. Oxford, Oxford University Department of Educational Studies.

Woolnough, B. E. (1991b) *Practical Science*. Buckingham, Open University Press.

Woolnough, B. E. (1993) 'Teachers' perception of reasons students choose for, or against, science and engineering', *School Science Review*.

Woolnough, B. E. (1994) 'Factors affecting students' choice of science and engineering', *International Journal of Science Education*.

Woolnough, B. E. and Allsop, R. T. (1985) *Practical Work in Science*. Cambridge, Cambridge University Press.

Woolnough, B. E. and Toh, K. A. (1990) 'Alternative approaches to assessment of practical work in science', *School Science Review*, 71(256): 127–31.

Ziman, J. (1972) 'Puzzles, problems and enigmas' BBC Broadcast reprinted in 1981 by Cambridge University Press.

Index

PRACTICAL SCIENCE
THE ROLE AND REALITY OF PRACTICAL WORK IN SCHOOL SCIENCE

Brian Woolnough (ed.)

Science teaching is essentially a practical activity, with a long tradition of pupil experimental work in schools. And yet, there are still large and fundamental questions about its most appropriate role and the reality of what is actually achieved. What is the purpose of doing practical work? – to increase theoretical understanding or to develop practical competencies? What does it mean to be good at doing science? Do we have a valid model for genuine scientific activity? – and if so do we develop it by teaching the component skills or by giving experience in doing whole investigations? What is the relationship between theoretical understanding and practical performance? How significant is the tacit knowledge of the student, and the scientist, in achieving success in tackling a scientific problem? How important are such factors as motivation and commitment? What do we mean by transferability and progression in respect to practical work? – do they exist? – can they be defined? How can we assess a student's practical ability in a way which is valid and reliable and at the same time encourages, rather than destroys, good scientific practice in schools? This book addresses such questions.

By bringing together the latest insights and research findings from many of the world's leading science educators, new perspectives and guidelines are developed. It provides a re-affirmation of the vital importance of practical activity in science, centred on problem-solving investigations. It advocates the need for students to engage in whole practical tasks, in which all aspects of knowledge (tacit as well as explicit), of practical ability, and of personal attributes of commitment and creativity, are interacting. While considering the particularly pertinent issues arising from the National Curriculum for Science in England, its discussion is equally germane to all concerned with developing good practical work in schools.

Contents

Setting the scene – Practical work in school science: an analysis of current practice – The centrality of practical work in the Science/Technology/Society movement – Practical science in low-income countries – a means to an end: the role of processes in science education – Practical work in science: a task-based approach? – Reconstructing theory from practical experience – Episodes, and the purpose and conduct of practical work – Factors affecting success in science investigations – School laboratory life – Gender differences in pupils' reactions to practical work – Simulation and laboratory practical activity – Tackling technological tasks – Principles of practical assessment – Assessment and evaluation in the science laboratory – Practical science as a holistic activity – References – Index.

Contributors

Terry Allsop, Bob Fairbrother, Geoffrey J. Giddings, Richard Gott, Richard F. Gunstone, Avi Hofstein, Richard Kimbell, Vincent Lunetta, Judith Mashiter, Robin Millar, Patricia Murphy, Joan Solomon, Pinchas Tamir, Kok-Aun Toh, Richard T. White, Brian E. Woolnough, Robert E. Yager.

224pp 0 335 09389 2 (Paperback) 0 335 09390 6 (Hardback)

BIOTECHNOLOGY IN SCHOOLS
A HANDBOOK FOR TEACHERS

Jenny Henderson and Stephen Knutton

In recent years there has been spectacular growth in biotechnology and in its importance for the school curriculum. This handbook offers teachers:

- an overview of the significance and scope of biotechnology
- an introduction to the content of biotechnology and its relevance to the everyday world
- a guide to how biotechnology fits into the National Curriculum, within and across subject disciplines
- appropriate teaching strategies
- suggestions for practical work
- case studies and other material which can be used directly with sixth form students
- a glossary of terms
- a guide to resources
- coverage of safety issues.

This is an essential resource for practising and trainee teachers of science and technology.

Contents
What is biotechnology? – Biotechnology and the school curriculum – Biotechnology and the food industry – Biotechnology and medicine – Biotechnology in agriculture – Biotechnology and the environment – Biotechnology, fuels and chemicals – Biotechnology through problem solving – Biotechnology through discussion-based learning – Practical considerations – Resources – Glossary – Appendix – References – Index.

176pp   0 335 09368 X (Paperback)   0 335 09369 8 (Hardback)

LEARNING AND TEACHING IN SCHOOL SCIENCE
PRACTICAL ALTERNATIVES

Di Bentley and Mike Watts

This book provides a series of different approaches to teaching school science. These approaches will be of use not only to science teachers but also to teachers outside science and in different parts of the education system.

The book is organized as follows. The first chapter looks at pressures for change: the authors show that science teachers need to adopt new and different approaches to teaching and learning. In particular, the authors focus on the notion of active learning – a theme that runs through the remainder of the book. In the following chapters, case studies are clustered around a series of themes. The final chapter summarizes the approaches and their implications for teaching science for the National Curriculum.

In general, the book is a useful, practical guide to a variety of strategies and classroom activities: a collection of experience and ideas about different teaching methods which will benefit both trainee and practising teachers. It will appeal to those engaged in initial training and in-service work, as well as to teachers who are keen to innovate.

Contents

The Contributors

Brigid Bubel, Bev A. Cussans, Margaret Davies, Rod Dicker, Mary Doherty, Hamish Fyfe, John Heaney, Martin Hollins, Joseph Hornsby, Andy Howlett, Pauline Hoyle, Harry Moore, Robin Moss, Phil Munson, Philip Naylor, Jon Nixon, Mick Nott, Anita Pride, Peter Richardson, Linda Scott, Brian Taylor, David Wallwork, Norma White, Steve Whitworth.

224pp   0 335 09513 5 (Paperback)   0 335 09514 3 (Hardback)